The Last Call:

How Climate Change Redefines Our Future

Francisco José Hurtado Mayén

Disclaimer

This book is inspired by current climate events and is based on information and data available as of the date of publication. However, due to the dynamic and ever-evolving nature of climate change and its effects, some data and statistics may have changed or evolved since its collection.

The author has made every effort to present the information in a precise and up-to-date manner, but the accuracy or completeness of all contents is not guaranteed. Readers are advised to consult official and up-to-date sources to stay informed about climate change and its implications.

This book aims to encourage reflection and dialogue around climate action and its impact on our lives and communities, and should not be construed as a comprehensive technical or scientific document.

November 2024

CONTENT

WE URGENTLY NEED TO ACT

In recent years, extreme weather events have become more frequent and devastating, leaving deep traces in the lives of those who suffer from them. Spain has witnessed increasingly intense phenomena, such as the recent DANA that severely affected regions such as Valencia, Albacete and Andalusia. These events not only involve torrential rains and strong winds; They bring with them a devastating impact on everyday life, from the loss of homes and livelihoods to the collapse of critical infrastructure.

This year's DANA is a clear example of how climate change is modifying our reality. In Valencia, water covered entire streets, and numerous families had to be evacuated while their homes were completely flooded. Property damage is not the only challenge, but also the emotional and psychological toll for those who lose everything in a matter of hours. As rescue teams worked to save lives and recover what was lost, those affected tried to process the impact of a catastrophe they never imagined.

In Albacete, rural and agricultural areas suffered particularly. Razed crop fields, blocked roads and destroyed irrigation systems have put in check agricultural production, on which so many families and businesses in the region depend. Crop failure not only affects farmers, but also the entire food chain, driving up prices and creating shortages. The ripple effect of these floods shows how the impact of an extreme event can reverberate throughout the economy and disrupt the lives of millions of people, even if they don't experience it directly.

In Andalusia, the storms affected residential areas and small businesses. Entire neighborhoods were isolated, and many residents were forced to leave their homes, taking only the essentials. For some family-owned businesses, which were already struggling to recover from the pandemic, these damages represent an insurmountable economic burden. Lack of electricity, service cuts and the destruction of roads and bridges pose a challenge to communities, which depend on safe infrastructure for daily life and economic development.

These climatic phenomena not only devastate material goods, but also profoundly alter the stability and well-being of communities. The disruption of basic infrastructure, such as transport, electricity and drinking water networks, directly affects the quality of life, access to resources and the functioning of health and education institutions. Every disaster brings with it a sequel of problems that require an urgent and effective response.

Events like this year's DANA are a warning of what's to come if we don't take urgent action on climate change.

The frequency and intensity of these phenomena increase as the planet warms, and this is just the beginning of a larger problem. Recognizing how these disasters affect daily life and are intertwined with the economy and social stability is crucial to understanding the magnitude of the climate crisis.

Stories of people affected

Behind every natural disaster are faces, stories, and lives that change forever. Extreme weather events are not just news, but traumatic experiences that leave deep scars on those who suffer from them. Below, we share some testimonies of people who have suffered first-hand the impact of the recent DANA in Spain. Their stories invite us to put ourselves in their shoes and reflect on the vulnerability of our lives in the face of the unstoppable force of nature.

Rosa and Antonio: "Everything we built in 30 years was washed away by water in one night"

Rosa and Antonio had been living for more than three decades in a house that they had built themselves in a small town in Valencia. With effort and dedication, they had transformed that house into a home full of memories and family projects. However, in a single night, the DANA flooded the house up to the roof. "We couldn't save anything," Rosa says through tears. "We woke up with the water rising quickly, and the only thing we had time to do was run away with the clothes we were wearing."

Antonio remembers the devastation when he returned: "It wasn't just water; It was mud, shattered furniture, walls that seemed to have collapsed. Everything we built in 30 years was washed away by water in one night. The feeling of helplessness is indescribable." For them, the loss is not only material; it is the emotional weight of knowing that everything that symbolized his history and his work has disappeared.

Carmen: "I don't know if I'll be able to start over"

In Albacete, Carmen, a 52-year-old farmer, lives with the uncertainty of not knowing if she will be able to continue with her olive farm. The DANA wiped out their crops, destroying this year's entire harvest and leaving the soil saturated with water. "For me, farming is more than a job. It's my family's legacy, what my parents and grandparents taught me," she explains. "Now I see the flooded field and I think of all that we have lost."

Carmen is facing significant economic losses that, in her case, are not fully covered by insurance. "Every year it gets harder," she says sadly. "The storms are stronger, the droughts longer. I don't know if I will be able to start over.

Maybe it's time to give up on this dream and accept that nature no longer allows us to go on with the life we knew."

Lucas and Sara: "We lost everything in the family business"

Lucas and Sara, a young couple from Andalusia, had opened a small family restaurant two years ago. Investing in their own business was a dream come true, a project to which they dedicated their savings and a lot of time. However, the DANA flooded the premises, destroying the appliances, furniture and all the merchandise in the warehouse. "We arrived at the restaurant after the storm and we couldn't believe our eyes," Sara says. "There was water and mud everywhere, it seemed as if someone had destroyed every corner."

Lucas adds: "We are aware that there will be aid, but sometimes that is not enough. The cost of rebuilding everything is enormous and, after the pandemic, we do not have the resources to recover from another blow so hard. We ask ourselves every day if we will be able to move forward or if we will have to close permanently."

José: "Seeing how my neighborhood turns into a river is something I will never forget"

José, a resident of a neighborhood on the outskirts of Valencia, remembers how the streets he walked daily became large rivers during the DANA. "Hearing the sirens, seeing neighbors evacuating their homes, some with small children in their arms, is something I will never forget," she says. His home was partially damaged, and although he was able to save most of his belongings, the experience left a deep emotional mark on him. "I have never felt so much fear in my life. Seeing my neighborhood turn into a river in a matter of minutes makes you realize how fragile everything is."

Since then, José has lived with the anxiety that another similar event could happen at any moment. "Every time heavy rains are announced, my wife and I are on alert. We no longer sleep peacefully. This situation has changed us forever."

These testimonies reflect the human and emotional toll of climate change. The loss of material goods is devastating, but even more so is the loss of security, hope and a hard-built life. The experiences of Rosa, Antonio, Carmen, Lucas, Sara and José not only show us the suffering of those who live through these tragedies, but also invite us to reflect on our responsibility to act to prevent stories like these from repeating themselves.

These personal accounts remind us that climate change is not a distant problem; it is a threat that is disrupting lives and entire communities. The emotional connection we feel when reading these stories should be the engine that drives us to make conscious decisions, for the benefit of those who have suffered and the generations that are yet to come.

Climate as an everyday reality

Climate change is no longer a future concern or a problem reserved for future generations. Today, its effects have taken hold in our daily lives and are shaping both our decisions and our habits and lifestyles. Extreme weather and its consequences have become part of our reality, affecting millions of people in ways we would not have imagined before.

One of the most subtle but profound consequences of this crisis is the psychological impact it generates. Insecurity and fear in the face of extreme weather events, such as the recent storms in Spain, create a sense of vulnerability that affects the emotional stability of those who experience them directly, and also of those who fear being next. This phenomenon, known as "climate anxiety", particularly affects young people, who see climate change as a threat to their future. The uncertainty they feel when imagining a future in which these events intensify and become more frequent affects their quality of life and has begun to be recognized as a direct consequence of the climate crisis. This anxiety and psychological stress reflect an emotional burden that grows in parallel with the challenges posed by climate change.

Climate change is also transforming our habits in tangible ways. In many regions, people are forced to modify their daily routine to adapt to this new normal. In flood-prone areas, for example, residents have learned to take precautions that were not previously necessary, such as keeping documents and valuables in high places or preparing for quick evacuations. In areas suffering from increasingly extreme heat waves, everyday activities such as working, exercising or even sleeping adapt to temperatures that make the environment hostile. This forced adaptation is reflected in the growing demand for services and products that mitigate the impact of climate, from air conditioning systems in places where they were not previously considered essential, to insurance against natural disasters in areas that did not usually need them. This change in habits is not a choice, but a measure of survival in the face of an increasingly unpredictable and hostile environment.

In addition, the impact of climate change goes far beyond material losses and profoundly affects the social fabric. Communities affected by natural disasters not only lose basic infrastructure, such as roads, hospitals, and schools, but also face the complex task of rebuilding their lives and sense of belonging. In many rural or economically disadvantaged areas, where resources for recovery are limited, the aftermath of these events lasts for years, affecting the development and well-being of their inhabitants. A lack of resources not only slows down recovery, but also increases the risk that these communities will be pushed into climate displacement, forcing people to leave their homes in search of better conditions. This forced migration redefines the concept of home and belonging, creating new social dynamics in a world where climate change imposes its own rules.

Today, climate change is a driving force shaping the way we live, work, and plan for our future. From the organization of our cities to the decisions we make on a day-to-day basis, extreme weather has become a factor that influences almost every aspect of our lives. The human and social cost of this crisis is immense and affects people of all ages and circumstances. We live in a world where climate is no longer a stable variable, but an uncontrollable force that can transform in an instant what we know and appreciate. Becoming aware of this reality is the first step to face it; Understanding how climate change affects every aspect of our lives allows us to see the urgency of the situation and the need to take concrete action to protect what we can still preserve.

Climate change is our current reality. Adapting and responding to this crisis is our only option to protect the present and ensure a future for generations to come.

The Scientific Evidence

Climate change is a phenomenon that has gained great relevance in recent decades due to its visible and growing effects on the environment. To understand it better, it is essential to break down some key concepts. In simple terms, climate change refers to significant and long-lasting variations in global weather patterns. Although the Earth's climate has changed naturally over millions of years, today we are facing accelerated change exacerbated by human activity, making it an urgent concern.

One of the most relevant factors in this process is global warming, which refers to the sustained increase in the average temperature of the planet. This warming is a direct consequence of the accumulation of certain gases in the atmosphere, known as greenhouse gases. These gases, including carbon dioxide (CO_2) and methane (CH_4), have the ability to trap the heat radiated by the Earth, preventing it from escaping into space and generating a "blanket" effect that warms the atmosphere.

CO_2 is one of the main gases responsible for global warming and comes largely from the burning of fossil fuels, such as oil, gas, and coal, in industrial activities, transportation, and electricity generation. This gas can remain in the atmosphere for centuries, accumulating and progressively increasing in concentration, which intensifies the greenhouse effect. On the other hand, methane,

19

although found in lower concentrations, is significantly more potent than CO_2 in terms of its ability to trap heat. This gas is mainly emitted through agricultural activities (such as livestock and rice cultivation), waste management and fossil fuel extraction.

The greenhouse effect is a natural process that allows the Earth to maintain an adequate temperature to support life. However, human activity has increased the concentrations of these gases to unsustainable levels, intensifying the greenhouse effect and causing global warming. As the planet's temperature rises, the consequences become apparent: melting glaciers, rising sea levels, extreme weather events, and changes in ecosystems that put countless species, including our own, at risk.

Understanding these concepts and their relationship allows us to better understand the urgency of reducing greenhouse gas emissions and adopting more sustainable practices. Climate change is a complex problem, but its explanation can be simplified: it is ultimately the result of an imbalance in the Earth's natural system caused by our activities.

Human activity and climate change

The link between human activity and climate change is undeniable and has been widely supported by the scientific community in recent decades. Human practices in key sectors such as industry, transport, intensive agriculture and deforestation have intensified global warming and contributed to the increase in extreme weather events, creating a devastating impact on the global climate.

Industry is one of the biggest contributors to carbon dioxide (CO_2) emissions and other greenhouse gases. Burning fossil fuels to produce electricity and power in factories and industrial plants releases massive amounts of CO_2, which is trapped in the atmosphere. According to data from the *Global Carbon Project*, in 2020 global CO_2 emissions reached 34.1 billion tons, mainly driven by the energy and industrial sector. This accumulation of gases intensifies the greenhouse effect, warming the atmosphere and altering global weather patterns.

The transportation sector is another important factor, as the use of vehicles running on fossil fuels contributes significantly to greenhouse gas emissions. According to the *International Energy Agency (IEA)*, transportation was responsible for 24% of global CO_2 emissions in 2019. Every time we use a car, a truck or an airplane, we are generating emissions that accelerate climate change. With the growing demand for mobility, emissions from this sector have increased exponentially, especially in urban areas.

Intensive agriculture also plays a crucial role in this climate crisis. Activities such as livestock farming generate large amounts of methane (CH_4), a greenhouse gas that, although less abundant than CO_2, has a much higher warming potential. Livestock production accounts for approximately 14.5% of greenhouse gas emissions, according *to the Food and Agriculture Organization of the United Nations (FAO)*. In addition, intensive agriculture employs nitrogen-rich fertilizers that release nitrogen oxides, another potent greenhouse gas. These processes not only contribute to global warming, but also alter ecosystems and cause biodiversity loss.

Deforestation is another significant driver of climate change. Forests function as carbon sinks, i.e. they absorb and store large amounts of CO_2. However, indiscriminate logging to make way for crops, grasslands, and urbanization reduces the Earth's ability to absorb carbon. According to a study published in *Nature Communications*, tropical deforestation is responsible for approximately 10% of global CO_2 emissions. Forest loss not only releases the carbon stored in trees, but also destroys natural habitats and puts thousands of species that depend on them at risk.

The effects of these activities are not just theoretical; we are already seeing their consequences in the form of increasingly frequent and intense extreme events. A study by *Climate Central* shows that heatwaves in Europe are now

10 times more likely than a century ago due to human-induced climate change. Extreme events such as hurricanes, wildfires, and droughts have become common, affecting millions of people around the world, causing material losses, displacement, and affecting food security and access to water.

This link between human activity and climate change leaves us with a clear message: our way of life is contributing to an unprecedented imbalance in the Earth's climate system. Recent figures and scientific studies confirm this, underlining the urgent need to reduce emissions and transform our production, transport and consumption patterns. Immediate and sustained action is essential to curb this crisis and build a sustainable future.

The short- and long-term effects

The effects of climate change manifest themselves on two temporal levels: in the short term, with extreme weather events that are already affecting our lives, and in the long term, with profound and potentially irreversible changes that could transform the planet as we know it. In the short term, we are experiencing increasingly intense and frequent floods, heat waves, and storms. These climate alterations have an immediate impact on communities around the world, generating material losses, displacement, disease and, in many cases, loss of human lives. According to the World *Meteorological Organization (WMO)* report, the frequency of extreme heat waves has increased dramatically in recent decades, and they are expected to continue to be more intense and prolonged. In Europe, heatwaves like the one in 2019 are now five times more likely due to climate change.

Intense floods and storms, such as those caused by the recent DANA in the Mediterranean, are becoming more common. Extreme precipitation is driven by rising ocean temperatures, which evaporate large amounts of water into the atmosphere, causing torrential rainfall. This phenomenon especially affects regions such as the Mediterranean, where heavy rainfall events in vulnerable areas such as coasts and densely populated urban areas are putting thousands of people at risk. The economic losses from these disasters are considerable; the *European Environment Agency* estimates that economic losses caused by climate-related extreme events in Europe reached €400 billion between 1980 and 2020.

In the long term, the effects of climate change can be even more devastating if immediate action is not taken. One of the biggest risks is rising sea levels, which threaten millions of people living in coastal areas. As global temperatures continue to rise, melting glaciers and thermal expansion of the oceans will raise sea levels, endangering coastal cities around the world. According to the *Intergovernmental Panel on Climate Change (IPCC),* sea levels could rise by 0.6 to 1.1 meters by 2100 if greenhouse gas emissions are not drastically reduced. This would have catastrophic consequences for areas such as the Nile Delta, South Asia and, in Europe, low-lying areas of the Netherlands and the Mediterranean coast, which could suffer permanent flooding.

Another long-term effect is the loss of biodiversity. Ecosystems are designed to live in specific climatic conditions, and the change in temperature and water availability is severely affecting many species. In the Mediterranean, an area rich in biodiversity, climate change is expected to lead to the disappearance of essential habitats and the extinction of numerous species. The *International Union for Conservation of Nature (IUCN)* warns that up to 25% of species in the Mediterranean are at risk of disappearing in the coming decades if temperatures continue to rise. Biodiversity loss is not only an ecological tragedy, but it also impacts agricultural systems, human health, and the livelihoods of millions of people.

Scientific projections make it clear that the long-term consequences of climate change can be catastrophic if action is not taken immediately. In a scenario of inaction, we will face a planet with unstable climates, scarce natural resources, mass displacement, and collapsed ecological systems. The need to act is not only based on a concern for the environment, but on the urgency of protecting human life, economies and the future of future generations. Current research and projections emphasize the importance of implementing large-scale solutions now, before these long-term effects become an inevitable reality. Acting quickly is crucial to prevent these immediate and future impacts from jeopardizing the stability of our planet.

Nature's Call

Nature is sending us increasingly intense and frequent signals that reflect the magnitude of the climate crisis. The extreme events we are witnessing are not mere coincidences or temporal anomalies; they are direct warnings of a planet that has been pushed to the limit. Every unprecedented heat wave, every devastating storm, and every prolonged drought are clear messages that our relationship with nature is broken. These signs are, in fact, "wake-up calls" that invite us to reflect and recognize the urgency of a profound change in the way we live and relate to the environment.

The frequency and intensity of these phenomena have increased alarmingly in recent decades. Heat waves, for example, are now more common and severe, affecting millions of people and endangering human lives. In places where extreme heat was once a rarity, record temperatures are now being experienced year after year. The World Meteorological Organization (WMO) has noted that the last five years have been the warmest on record, a trend that shows no signs of slowing down. These heat waves don't just cause discomfort; They have devastating effects on health, crops, and ecosystems, affecting every aspect of life.

Likewise, storms and hurricanes have gained in intensity, causing massive destruction and forced displacement around the world. Rising ocean temperatures provide more energy to these systems, generating torrential rains and hurricane-force winds that wipe out entire communities in a matter of hours. These catastrophes leave behind flooded cities, homeless families and an unsustainable economic burden for many countries. Nature is speaking in a language of destruction and emergency, and every storm becomes a warning of what is to come if we don't act.

The drought is another clear manifestation of this crisis. Once-fertile regions are becoming arid lands, unable to sustain crops and threatening the food security of millions of people. The lack of water, a fundamental resource for life, is a constant reminder of how our broken relationship with the planet is affecting natural cycles that used to be predictable and reliable. Agriculture, rural communities, and biodiversity are paying the price for this imbalance, and with each passing year, drought becomes a more persistent and complex problem.

Nature, through these extreme events, is speaking to us with an urgency that we cannot ignore. Our relationship with the environment has become unsustainable, and these "wake-up calls" are a reflection of a planet that is reacting to centuries of exploitation and neglect. We have broken the balance that allowed us to live in harmony with ecosystems, and now we are facing the consequences. Every extreme event is a reminder that the time to act is now, that we can no longer ignore the warning signs of a planet asking us for change.

These signs are an opportunity to reflect on our impact and take responsibility for turning the tide. Ignoring them would be a negligence that will not only affect our generation, but will endanger future ones. Nature is giving us one last chance to re-establish our relationship with it, to learn from these warnings and take urgent action that will allow us to live in balance with the planet.

Solidarity and Empathy with those affected

We want to express our sincerest solidarity and empathy with all the people who have suffered devastating losses due to the DANA and other extreme weather events that have recently hit various regions of our country and the world. We know that behind every flood, every storm and every heatwave are personal stories of grief, families who have lost their homes, their belongings and, in many cases, their livelihoods. This suffering is real and deserves to be recognized with all the sensitivity and respect that comes with seeing how nature, altered by climate change, affects the lives of thousands of people.

It is critical that, as a society, we come together in support of those who are facing the consequences of these disasters. Empathy should not only remain in words; It is necessary for each of us to contribute to reaching out to those affected, whether through donations, volunteering or simply offering our emotional support. In addition to immediate help, it is crucial that we also think about the future and how we can act to prevent similar tragedies from happening again. We cannot allow these types of events to become a "new normal" that accepts the suffering of so many people as inevitable.

Acknowledging the suffering of those affected and coming together in solidarity is the first step in building a strong and compassionate social response. At the same time, we must understand that our responsibility does not end with help; It also involves making a commitment to the change needed to reduce the impact of climate change. If we do not act now, we will be condemning future generations to live in a world where these extreme events will become even more frequent and devastating. May this message of solidarity serve not only as comfort to those who have lost so much, but also as a call to collective action to protect each other and the planet we share.

An invitation to reflection and action

This is the time to stop and reflect deeply on our role in the climate crisis. Each of us, both as individuals and as members of a global society, has an inescapable responsibility for the protection and preservation of the planet. We cannot go about our daily routines without questioning the impact they have on the environment, nor can we ignore the consequences of our consumption decisions, our habits and our priorities. The magnitude of the climate crisis demands that each of us look inward and ask ourselves: what can I do to reduce my footprint on the planet?

The answer is not simple, but it starts with awareness. Being aware of the seriousness of the situation and our capacity for change is the first step. We can take small, everyday actions, such as reducing the use of plastics, opting for sustainable means of transport, saving energy, and supporting local and sustainable products. However, this is just the beginning. It is also our responsibility to demand structural changes and support environmental policies that seek to protect ecosystems, curb pollution, and reduce greenhouse gas emissions. As citizens, we have the power to influence the decisions of those who govern us and the direction our communities and cities take.

It is critical to take an active role in this fight and understand that change will not happen if we expect others to act for us. The climate crisis concerns us all and its solution depends on a collective effort. Every action, no matter how small it may seem, adds up and has the potential to inspire others to act. We cannot leave this burden to future generations; It is our responsibility to protect the planet and ensure that generations to come find a world in which they can live and thrive.

The time to act is now. We can no longer postpone our responsibilities or pass the burden on to the unborn. Now is the time to come together, take concrete action, and commit to a sustainable future. For ourselves, for our families, and for all generations to come, we must act with determination and urgency. The planet is our home, and protecting it is our non-negotiable duty.

INVISIBLE COSTS OF CLIMATE CHANGE

Climate change is generating profound economic repercussions and transforming key sectors of the economy in Europe and especially in Spain. The consequences affect everything from agriculture to tourism, as well as infrastructure and industry, and are generating economic losses and putting many jobs at risk. The effects are explored in detail below:

Agriculture and Food Security

The agricultural sector is one of the most affected by climate change due to its direct dependence on weather conditions. In Spain, a country with extensive agricultural areas and an increasingly extreme Mediterranean climate, the change in temperature and precipitation patterns is severely impacting production. Prolonged heatwaves and recurrent droughts reduce harvests of staple crops such as wheat, maize and olives, all of which are essential for both domestic consumption and export.

Olive production, in particular, is vulnerable to high temperatures, which not only affect the quantity of fruit, but also its quality, with repercussions on the olive oil industry, one of Spain's most emblematic products. Likewise, crops such as vines, which withstand specific climate conditions, are seeing their productions decrease. A recent report by the

Ministry of Agriculture, Fisheries and Food indicates that losses in agricultural production due to extreme events in the last five years amount to hundreds of millions of euros. In addition, the effects on agriculture generate a chain reaction that affects the food supply chain and raises food prices, which increases the cost of living and affects the food security of the population.

The impacts of climate change on agriculture also put rural jobs and the viability of smallholder farms at risk, which cannot always cope with the losses or additional costs needed to adapt, such as installing efficient irrigation systems or using heat-resistant crop varieties. The loss of these jobs and the abandonment of agricultural land affect local economies and favour the depopulation of rural areas, a phenomenon that is already worrying in several regions of Spain.

Tourism in decline

Tourism is another sector vulnerable to climate change, especially in southern Europe, where it represents a key source of income and jobs. Mediterranean regions, which have traditionally attracted millions of tourists thanks to their warm climate and natural attractions, are experiencing a negative impact due to rising temperatures, the risk of wildfires, and environmental degradation.

The extreme heat, which is becoming more common during the summer months, causes many visitors to reconsider their destinations and travel seasons, opting for cooler places or visiting during less warm seasons. This change is particularly affecting beach destinations in Spain and other Mediterranean countries, where temperatures can easily exceed 40°C in summer, putting the comfort and health of tourists at risk. In addition, wildfires, which become more intense every year, destroy natural areas that are often major tourist attractions and force the closure of recreational areas and hiking trails, generating losses for both local businesses and seasonal workers.

The loss of biodiversity in natural ecosystems, such as national parks and marine reserves, also reduces tourist attractiveness, as environmental degradation decreases the quality of the experience for visitors. According to data from the National Institute of Statistics (INE), tourism in areas especially vulnerable to climate change could be reduced by 20% in the next two decades if adaptive measures are not implemented, which would have a direct impact on income and jobs in the sector.

Vulnerable Infrastructure

Infrastructure across Europe, from roads to bridges to power grids, is being subjected to extreme conditions for which it was not designed. Floods and heatwaves increase the wear and tear on key infrastructure, which raises maintenance and repair costs and affects the economy and the daily lives of the population.

Flooding, increasingly common in urban areas, causes damage to roads and transportation systems, and can disrupt power supply and access to clean water. These disruptions affect not only the lives of citizens, but also the operation of businesses, which depend on these infrastructures to operate normally. Heat waves, on the other hand, can cause deformations in asphalt and overheating in power grids, increasing the risk of blackouts and problems in public transport.

Coastal infrastructure, in particular, is vulnerable to rising sea levels and more intense storms. In cities like Valencia and Barcelona, where port and coastal infrastructure is crucial for trade and tourism, the impact of climate change is already becoming apparent. Coastal cities are forced to invest in protective measures, such as dikes and barriers, to adapt to these new conditions, which represents a significant cost for local and national governments.

Industry Transformation

The industry is also under pressure to adapt to sustainability demands and reduce its environmental impact. However, the transition to sustainable practices and the modernization of processes represent a high cost, especially for small and medium-sized companies that do not have sufficient resources to implement profound changes. Industries that depend on natural resources, such as energy, mining, and manufacturing, are particularly exposed to climate change, as resource depletion and rising temperatures affect their production processes.

Companies that fail to modernise and adapt may lose competitiveness, especially in a context of regulatory change, where European governments are setting increasingly stringent restrictions on emissions and sustainability. The transformation of industry is essential to meet greenhouse gas emission reduction targets, but the transition process also puts jobs in traditional sectors at risk and may generate instability in some local economies.

Despite the challenges, several industries are taking steps towards sustainability. The energy sector, for example, is investing in renewables, such as solar and wind, to reduce its dependence on fossil fuels. The automotive industry is also developing electric vehicles and clean technologies in response to market demands and climate policies. However, for these transformations to become widespread and reach their maximum impact, continued support from both government and society as a whole is necessary.

These areas show how climate change profoundly affects the economy and jobs in Europe and Spain, from the countryside to the city, and from tourism to industry. The economic impact goes beyond immediate material losses, threatening the stability of entire sectors and putting thousands of jobs at risk. Adapting to this new reality requires coordinated and urgent action, with investments and policies that protect both the affected sectors and the people whose lives depend on them.

Public Health at Risk

Climate change not only affects the environment and the economy; Its effects also have serious implications for public health. As temperatures rise and weather patterns become more extreme, conditions are conducive to the proliferation of diseases, deterioration of air quality and heat stress. In addition, changing cropping patterns and food availability pose new challenges in nutrition and food security. This section explores how these effects impact public health and why it is essential to adapt to this new reality to protect the well-being of the population.

Respiratory Diseases and Air Quality

Climate change and pollution are contributing to deteriorating air quality, increasing cases of respiratory diseases such as asthma, bronchitis and other lung conditions. As temperatures rise, emissions of pollutants and allergens also intensify. In densely populated urban areas, where air pollution is higher due to traffic and industrial activities, levels of fine particulate matter and other pollutants are especially harmful to health. The World Health Organization (WHO) warns that long-term exposure to these pollutants is responsible for millions of premature deaths each year and significantly increases the risk of respiratory and cardiovascular diseases.

Wildfires, exacerbated by climate change, are also generating large amounts of airborne particulate matter that affects respiratory health. Heat waves and drought create favorable conditions for the spread of fires, which release smoke and dangerous pollutants that can be transported over long distances, affecting both nearby populations and those in remote regions. Inhaling fine particles can aggravate asthma and lead to severe respiratory problems, especially in children, the elderly, and people with pre-existing respiratory conditions. As these fires become more frequent and severe, the risk to public health continues to increase.

Insect-Borne Diseases

Global warming is allowing the geographic expansion of disease-carrying insects, such as mosquitoes and ticks, into regions where they were not common before. Diseases such as dengue, Zika and West Nile virus, traditionally present in tropical areas, are appearing in Europe, posing an emerging public health challenge. Changing temperature and precipitation patterns have created suitable conditions for the reproduction of these vectors in areas where previously the climate was too cold for their survival.

In Spain, for example, cases of dengue and West Nile virus have already been reported in areas where these diseases were unthinkable a few decades ago. The *European Centre for Disease Prevention and Control* (ECDC) estimates that the incidence of mosquito-borne diseases in Europe could increase dramatically in the coming years if current warming trends continue. The expansion of these vectors poses an urgent need to adapt health systems and improve epidemiological surveillance in Europe. Health systems must be prepared for the prevention, detection and treatment of diseases that previously did not represent a threat, and that now pose a real risk to the population.

Heat Stress and Heat Mortality

Extreme heatwaves, which are becoming more frequent and prolonged, present a significant health risk, especially among older people and those with chronic diseases. Prolonged exposure to high temperatures can lead to heat stress, heat exhaustion, and even heat stroke that, if not treated immediately, can be fatal. Heat waves also increase mortality related to cardiovascular and respiratory diseases, as the human body, in trying to adapt to high temperatures, experiences an additional burden that can aggravate these conditions.

Studies show that climate change has increased the frequency and duration of heatwaves around the world, and Europe is no exception. During the 2003 heatwave, which was one of the deadliest in Europe's recent history, it is estimated that more than 70,000 people died across the continent. Projections indicate that, without effective climate action, these types of extreme events will become increasingly common. Public health systems must prepare to meet these challenges by implementing prevention and response measures, such as emergency plans, cooling infrastructure, and awareness campaigns.

Food Insecurity and Malnutrition

Climate change is also affecting agricultural production and access to food, increasing the risk of food insecurity and malnutrition in diverse populations. Alterations in temperature and precipitation patterns make it difficult to grow food in many regions, which in turn affects the availability, quality, and price of commodities. The decrease in the production of foods such as cereals, fruit and vegetables in Spain and other parts of Europe due to droughts and heatwaves has a direct impact on the population's diet and food security.

Food insecurity not only affects people in terms of the quantity of food available, but also in its nutritional quality. Lack of access to a variety of fresh foods increases the risk of malnutrition and nutritional deficiencies, especially in vulnerable communities already facing economic hardship. The *Food and Agriculture Organization of the United Nations* (FAO) warns that climate change could significantly reduce the availability of essential foods in the coming decades, mainly affecting low-income populations and rural sectors.

Food insecurity also has a long-term health impact, as a diet lacking in essential nutrients can lead to growth problems in children, weaken the immune system, and increase susceptibility to chronic diseases. To meet these challenges, it is critical that governments and international organizations work on adaptation and resilience measures, such as the development of climate-resilient crops and the implementation of policies to support farmers.

The effects of climate change on public health are profound and varied, affecting everything from air quality to food security. These impacts not only place a burden on health systems, but also affect the quality of life and well-being of millions of people. Adaptation and preparedness are essential to protect the health of the population against these risks, and should be a priority in public health policies. Climate change is a global health crisis that requires immediate and sustainable responses to mitigate its effects and protect present and future generations.

Emotional and Psychological Cost

Climate change is not only transforming the physical environment and economy, but also the emotional lives of millions of people around the world. As evidence of environmental degradation accumulates and extreme weather events become more frequent, a less visible but equally real phenomenon is growing: the psychological impact of the climate crisis. This emotional toll ranges from anxiety about the future to trauma for those who have experienced natural disasters, and has profound effects on collective mental health. It then explores the different aspects of the psychological cost of climate change and how it affects both individuals and entire communities.

Climate Anxiety and Fear of the Future

"Climate anxiety" is a form of distress that arises from deep and persistent concern about the future of the planet and the effects of climate change. While this anxiety can affect people of all ages and backgrounds, it's especially prevalent among young people, who see their future become uncertain on an increasingly unstable planet. Climate anxiety manifests itself as a combination of fear, helplessness, and hopelessness, and affects not only on an emotional level, but also on the way these young people plan their lives, their studies, and their careers.

Many young people experience this anxiety in moments of introspection, such as when reading news about extreme weather events, observing the consequences of natural disasters, or reflecting on the lack of effective action by world leaders. This climate anxiety can become a paralyzing feeling, affecting their ability to enjoy the present and pushing them to question important decisions, such as having children or starting a family. For some, the idea of bringing new generations into a world threatened by climate change is distressing, highlighting the level of psychological impact this crisis can have on personal life choices.

Post-Traumatic Stress in Those Affected by Natural Disasters

People who directly experience the effects of extreme weather events, such as floods, wildfires, hurricanes, or storms, may develop post-traumatic stress disorder (PTSD) due to the intensity and impact of these events. This type of trauma is not only caused by the experience of the disaster itself, but also by the loss of homes, belongings and livelihoods, as well as by the forced displacement and effort required to rebuild their lives.

PTSD can manifest itself through nightmares, intrusive memories, extreme anxiety, and the inability to cope with situations reminiscent of the traumatic event. For example, those who have experienced a devastating flood may feel anxiety or fear every time it rains, fearing a new catastrophe. The psychological aftermath of these events can be long-lasting, and many affected people find it difficult to resume their lives as normal even years after the disaster. The traumatic experience affects both individuals and entire communities, and those who do not receive adequate psychological support may face persistent difficulties in their recovery process.

Public health systems are overwhelmed by the increased demand for psychological support in regions affected by natural disasters, underscoring the need to integrate mental health care as part of climate response and emergency plans. Providing affected people with the necessary emotional support is crucial to help them regain their balance and ability to face future challenges.

Impact on Collective Mental Health

Climate change generates a sense of vulnerability and powerlessness that goes beyond individuals and affects the mental health of society as a whole. The idea of living in an uncertain world, where extreme events affect an increasing number of people, has led to an increase in states of discouragement, apathy and hopelessness in large sectors of the population. This phenomenon, known as "eco-anxiety", especially affects those who are highly informed about the climate crisis and feel that their individual actions are not enough to curb a problem of such magnitude.

Collective mental health is affected by a perception of constant threat that generates anguish and demotivation. For many people, the lack of climate action at the global level generates a feeling of helplessness that translates into inaction and emotional disconnection. In vulnerable communities, where the effects of climate change are already palpable, this deteriorating collective mental health hinders social development and people's ability to mobilize for sustainable solutions. While climate concern may spur some to action, for others it generates an emotional charge that leaves them without energy to get involved, creating a cycle of frustration and resignation.

Psychological Adaptation and Resilience

Despite these negative effects, some individuals and communities develop resilience and a remarkable capacity to adapt in the face of the climate crisis. Resilience, understood as the ability to face and overcome adversity, is a key response that allows people to better manage anxiety and find a sense of purpose in the fight against climate change. In many communities affected by natural disasters, mutual support and solidarity become engines of resilience that allow individuals and groups to rebuild their lives and adapt to new conditions.

Fostering an action and adaptation mindset can help people channel climate anxiety into productive efforts. Participation in social movements, environmental advocacy organizations, and sustainability projects can provide individuals with a sense of purpose and empowerment. Studies have shown that those who engage in climate action have better mental health, as they feel they are actively contributing to a greater cause.

Emotional resilience can also be strengthened through education and awareness-raising, as understanding coping mechanisms and participating in disaster preparedness initiatives provides people with practical and emotional tools to cope with uncertainty. This approach helps to build a stronger and more conscious society, which is capable of facing climate challenges in a proactive and balanced way.

The emotional and psychological toll of climate change is real and increasingly apparent. From climate anxiety to post-traumatic stress to the impact on collective mental health, the climate crisis is affecting people of all ages and regions. Recognizing and addressing these effects is critical to building a comprehensive response to the crisis. Fostering resilience and psychological adaptation allows us not only to better face challenges, but also to find a sense of purpose in the fight against climate change, a cause that unites and motivates those who wish to protect the planet and ensure a safer future for generations to come.

NATURE AND US: BROKEN BALANCE

Ecological Footprint and Daily Activities

The ecological footprint is a fundamental concept to understand how our daily activities impact the planet. This indicator, which measures the use of natural resources and the generation of waste, is a key tool to assess the balance between our needs and the planet's capacity to regenerate. Below, the concepts of ecological footprint, its components and how they manifest themselves in our daily activities are explored in detail.

Definition and Calculation of the Ecological Footprint

The ecological footprint is an indicator that measures the environmental impact of our activities in terms of the resources we consume and the waste we generate. Basically, it refers to the amount of land and water needed to produce the goods and services we consume and to absorb the waste generated, particularly carbon dioxide (CO_2). This concept encompasses various aspects of our lifestyle, from food and energy consumption to waste generation and transport use.

There are several methods for calculating the ecological footprint, which can be adapted at different levels:

individual, community, business or even national. In general, the calculation considers factors such as:

- **Energy Consumption**: The electricity, gas, and other types of energy we use at home and in industry have a considerable impact on the ecological footprint. Each energy source has a different footprint, with fossil fuels being the most polluting.

- **Transportation Use**: The carbon footprint varies depending on the means of transportation we use. Private transport, especially petrol or diesel vehicles, generate more emissions than public transport, cycling or walking.

- **Food Consumption**: The food we consume has different ecological footprints depending on its origin and type of production. For example, animal products, such as beef, have a larger footprint due to the amount of resources required to produce them.

- **Consumption Habits and Waste**: The amount of products we buy and discard also influences our ecological footprint. The use of single-use plastics, the purchase of fast fashion clothing and the generation of non-recyclable waste increase the pressure on natural resources.

Calculating the ecological footprint is a useful tool for visualizing our impact and helps us understand what areas of our daily lives we can improve to reduce our burden on the planet.

How Our Daily Activities Contribute

Daily activities, although they may seem insignificant, add up and generate a considerable impact on the environment. Below are some common examples of how our lifestyle contributes to increasing our ecological footprint.

- **Use of Single-Use Plastics**: Plastic, especially single-use, is one of today's biggest environmental problems. Every time we use a plastic bag, bottle or wrap that we discard after a single use, we are contributing to the accumulation of waste that takes hundreds of years to decompose. In addition, plastic production depends on the petrochemical industry, which consumes large amounts of energy and generates CO_2 emissions.

- **Consumption of Meat and Animal Products**: The livestock industry is one of the main responsible for greenhouse gas emissions and the excessive consumption of water and soil. Beef production, in particular, has a high ecological footprint due to deforestation to create pastures and methane released by livestock. Reducing meat consumption and opting for plant-based diets can help significantly reduce your ecological footprint.

- **Excessive Energy Use in the Home**: The electricity and heating we use in the home account for a considerable part of our ecological footprint. Every time we leave lights on, use appliances unnecessarily or do not optimize the use of heating, we are increasing our energy consumption and, therefore, our footprint. Opting for efficient appliances and adopting energy-saving habits can reduce this impact.

- **Private Transport**: The use of private cars is one of the main sources of CO_2 emissions. Gasoline and diesel vehicles produce high levels of pollution and contribute to global warming. Using public transport, cycling or walking are more sustainable alternatives that reduce the ecological footprint and help to decongest cities.

Lifestyle in developed countries tends to be particularly resource-intensive and produces a significantly larger ecological footprint than in other regions. The combination of high levels of consumption, use of private vehicles and dependence on animal products contributes to a lifestyle that puts a great strain on natural resources and generates large amounts of waste. Globally, this consumption model is unsustainable and requires changes to balance people's well-being with the planet's capacity for regeneration.

Comparison between Countries and the Limit of the Planet

The concept of "planet boundary" or "carrying capacity" refers to the amount of natural resources that the Earth can sustainably produce and regenerate each year. When the ecological footprint of a country or population exceeds this limit, an ecological deficit is created that puts the stability of ecosystems at risk. In other words, we are using more resources than the planet can replenish, which is unsustainable in the long term.

When comparing the ecological footprint between different countries, great inequalities are observed. For example, in countries such as the United States, Canada or some in Western Europe, the ecological footprint per capita is much higher than in developing countries. If the entire world population lived at the consumption level of these countries, we would need several planets to sustain the demand for resources. According to the Global Footprint Network, if humanity as a whole were to reach the average lifestyle of countries like the United States, it would take more than five planet Earths to maintain that level of consumption.

This comparison shows not only the unsustainability of the consumption model in some countries, but also the importance of reducing the ecological footprint worldwide. Achieving a global ecological footprint that stays within the limits of the planet requires significant changes in our consumption habits, in environmental policies, and in the way we use natural resources.

Reducing the ecological footprint is a shared responsibility, but it requires an effort both individually and collectively. Adopting more sustainable consumption habits, reducing the use of finite resources and opting for less polluting alternatives are necessary steps to restore balance with the planet. The ecological footprint reminds us that each of our actions has an impact, and that it is possible to live in a way that respects the Earth's natural limits.

Dependence on Finite Resources

Dependence on finite resources is one of the greatest challenges facing humanity in its relationship with the planet. Many of the resources we use daily are non-renewable and are being extracted at an unsustainable rate, putting their availability for future generations at risk and threatening the ecosystems that depend on them. This section explores the nature of these finite resources, the crisis in water and energy availability, and the effects of intensive exploitation on ecosystems and local communities.

Finite Resources and the Extraction and Consumption Model

Finite resources are those that exist in limited quantities on the planet and cannot be renewed in the short term. These include oil, coal, natural gas, minerals, and certain types of wood. Unlike renewable resources, such as solar or wind, finite resources are not continuously regenerated, and once depleted, they cannot be recovered. The current economic model, based on intensive extraction and consumption, has created immense pressure on these natural reserves.

This model of extraction and consumption relies heavily on fossil fuels, which are not only limited, but also generate large amounts of greenhouse gas emissions when burned, contributing to climate change. In addition, the mining of minerals essential to modern technology, such as lithium, cobalt, and copper, has led to intensive exploitation that destroys ecosystems and threatens biodiversity. The demand for these resources continues to increase as the population and the global economy grow, leading to a situation of progressive depletion.

The depletion of finite resources is not only an environmental problem, but also an economic and social one. As resources become scarcer, the costs of extraction and production increase, making the products and services that depend on them more expensive. This situation also generates conflicts and tensions between countries that compete for access to these resources, especially in areas rich in minerals and oil. To ensure a sustainable future, it is essential that we move towards an economic model that promotes the use of renewable energies and adopts circular economy practices, where materials are reused instead of being discarded.

The Water and Energy Availability Crisis

Two of the most critical resources facing a serious crisis of availability are water and energy. The growing demand for water for agriculture, industry, and domestic consumption is leading to a shortage that affects millions of people around the world. Water, essential for life, is in a state of depletion in many regions due to heavy use, pollution and climate change. In areas where water is already scarce, competition between sectors and unequal access create significant social and economic problems.

In agriculture, which consumes about 70% of the planet's available freshwater, water scarcity is reducing production capacity and affecting food security. The industry also relies heavily on water for manufacturing and cooling processes, which adds pressure on water resources. In urban areas, lack of access to safe drinking water is a growing problem, and in many countries restrictions on use and high tariffs are being implemented to control demand. The water crisis not only affects human health and economic development, but also endangers aquatic ecosystems, where many species are at risk due to overexploitation and pollution of rivers and lakes.

On the energy side, reliance on fossil fuels such as oil, gas, and coal remains a reality in much of the world, despite the growth of renewables. The extraction, transport and consumption of these fuels not only release large amounts of CO_2, but also generate social and economic impacts. Volatility in oil and gas prices affects the global economy, while intensive extraction in remote and vulnerable areas often causes irreparable damage to ecosystems and local communities. The transition to renewable and clean energy sources, such as solar and wind, is critical to reducing dependence on fossil fuels and ensuring energy availability that is sustainable and accessible to all.

Impact of Exploitation on Ecosystems and Communities

Intensive extraction of natural resources has devastating consequences for ecosystems and the local communities that depend on them. Deforestation in tropical forests, for example, is a clear example of how resource exploitation can cause biodiversity loss and habitat destruction. Forests, which are essential for climate regulation and the water cycle, are being cleared for timber, create pastures for cattle and expand agriculture, especially for crops such as palm oil and soybeans. This deforestation not only threatens wildlife, but also reduces the ability of ecosystems to absorb carbon dioxide, accelerating climate change.

In the seas and oceans, overfishing is leading to the depletion of marine species and an alteration of aquatic ecosystems. Many fish species are on the verge of collapse due to intensive fishing and a lack of effective regulations. In addition, plastic and chemical pollution is affecting marine life and jeopardizing the food security of millions of people who depend on marine resources for their livelihoods.

Mining and oil rigs also exert significant pressure on vulnerable territories, such as rainforests and the habitats of indigenous communities. Open-pit mining and oil extraction not only destroy the soil and pollute water sources, but also generate social conflicts and displace populations that live in harmony with the environment. These intensive activities are often carried out without the consent of local communities, and the long-term impact on the environment and the well-being of these populations is enormous.

The exploitation of natural resources not only depletes ecosystems, but also creates inequalities and conflicts that affect the most vulnerable people. Indigenous and rural communities, who depend on natural resources for their daily lives, are the most affected by extractive activities. Protecting these territories and their resources is essential to ensure environmental justice and long-term sustainability.

Dependence on finite resources and intensive exploitation of them are driving the planet towards a point of depletion that threatens both biodiversity and social and economic stability. The shift towards a more sustainable consumption model, which promotes reuse and resource efficiency, is essential to reduce our dependence on finite resources and preserve ecosystems and the communities that depend on them.

Ecosystems in Climate Stability

Ecosystems play an essential role in regulating the climate and absorbing greenhouse gases. From vast forests to deep oceans to wetlands, these natural systems act as natural defenses against climate change. However, human activities have degraded and destroyed many of these ecosystems, reducing their capacity to absorb carbon and accelerate global warming. This section explores the role of ecosystems in climate regulation, the effects of their degradation, and solutions to restore their balance.

Ecosystems as Climate Regulators

Natural ecosystems, such as forests, oceans, and wetlands, act as key climate regulators. These environments have the capacity to absorb large amounts of carbon dioxide (CO_2), the main greenhouse gas responsible for global warming, and stabilize global temperatures. The specific role of some of these key ecosystems is described below:

- **Forests**: Forests, particularly rainforests, are known as "the lungs of the planet" because they absorb large amounts of CO_2 through photosynthesis. Trees and plants convert CO_2 into oxygen and store carbon in their biomass, such as trunks, branches, and roots. The world's forests are estimated to store approximately 861 gigatons of carbon, acting as sinks that help offset greenhouse gas emissions. In addition to absorbing CO_2, forests also play a crucial role in regulating precipitation patterns and protecting the soil from erosion.

- **Oceans**: The oceans absorb approximately 25% of the CO_2 emissions generated by human activities each year, making them one of the most important carbon sinks on the planet. CO_2 dissolves in seawater, where it is used by organisms such as phytoplankton to perform photosynthesis. In addition, the oceans help distribute heat globally through ocean currents, regulating the climate and maintaining stable temperatures. However, this process also generates a problem known as ocean acidification, as dissolved CO_2 changes the pH of the water, affecting the health of marine ecosystems.

- **Wetlands**: Wetlands, such as swamps and marshes, are highly effective ecosystems at capturing and storing carbon. Moist soils contain large amounts of decaying organic matter that, when covered in water, slowly decomposes and retains carbon for long periods of time. Peatlands, for example, store twice as much carbon as all the world's forests combined, even though they cover a much smaller area. In addition, wetlands contribute to the regulation of the water cycle and the reduction of floods, providing additional benefits for climate stability and biodiversity.

The Degradation and Destruction of Ecosystems

Deforestation, ocean pollution, expanding urbanization and intensive agriculture are rapidly degrading these natural ecosystems, decreasing their capacity to absorb CO_2 and affecting climate stability. Some of these issues are discussed in detail below:

- **Deforestation**: The indiscriminate clearing of forests for agricultural expansion, timber production and urban development is reducing the capacity of forests to act as carbon sinks. Every year, millions of hectares of forests are destroyed, especially in tropical regions such as the Amazon, which store huge amounts of carbon. When trees are cut down or burned, the carbon stored in their biomass is released into the atmosphere, contributing to greenhouse gas emissions and accelerating climate change.

- **Ocean Pollution**: Oceans face multiple threats due to pollution, such as oil spills, plastic waste, and excess nutrients from agricultural fertilizers. These pollutants affect the health of marine organisms, including phytoplankton, which plays a key role in CO_2 absorption. The degradation of coral reefs, which are home to extraordinary biodiversity, also reduces the ability of the oceans to support healthy ecosystems that contribute to climate regulation.

- **Urbanization and Intensive Agriculture**: The expansion of urban areas and intensive agriculture are altering the soil and eliminating natural habitats. Intensive agricultural practices, including the overuse of fertilizers and pesticides, degrade the soil and release nitrogen oxides, another greenhouse gas. The loss of natural areas, such as grasslands and wetlands, reduces the soil's capacity to store carbon and affects natural cycles, such as the water cycle, exacerbating the effects of climate change.

Conservation and Restoration as Solutions

To mitigate climate change and restore environmental balance, it is essential to conserve and restore ecosystems that act as carbon sinks and natural climate regulators. Conservation and restoration initiatives not only help reduce greenhouse gas emissions, but also protect biodiversity and strengthen the resilience of ecosystems to the impacts of climate change. Here are some key strategies and successful examples:

- **Reforestation and Forest Restoration**: Reforestation and restoration of degraded forested areas are effective solutions to increase carbon sequestration. Planting trees in areas where they have been cut down or destroyed not only helps absorb CO_2, but also recovers biodiversity and protects the soil from erosion. Initiatives such as the "Great Green Wall" in Africa seek to create a strip of vegetation in the Sahel to curb desertification and restore degraded lands, providing environmental and economic benefits for local communities.

- **Marine Conservation and Coral Reef Restoration**: Protecting marine areas and restoring coral reefs are critical to ocean health and their ability to sequester carbon. Marine conservation projects, such as marine protected areas (MPAs), limit human activity in certain areas, allowing ecosystems to recover and preserving marine species. Restoring coral reefs through transplants of resilient corals also contributes to maintaining marine biodiversity and protecting coastlines from erosion.

- **Protection and Restoration of Wetlands**: Wetlands are some of the most effective ecosystems in carbon sequestration and in protecting biodiversity. Restoring degraded wetlands and protecting peatlands are key strategies for mitigating climate change. In Canada, for example, peatland conservation programs have been implemented in the north of the country, which contribute significantly to carbon sequestration and help reduce the risk of wildfires.

Ecosystem conservation and restoration are powerful tools in the fight against climate change. These actions not only help reduce greenhouse gas emissions, but also provide additional benefits, such as protecting biodiversity, improving water quality, and reducing natural disasters. The implementation of policies and practices that promote the sustainability and resilience of ecosystems is essential to ensure a future in which nature and humanity can coexist in balance.

The Role of Society and Individuals

The climate and environmental crisis requires coordinated action at all levels, from the individual to the governmental and corporate spheres. To restore the planet's natural balance, we all have a role to play. From changing personal habits to demanding sustainability policies, rebuilding the natural balance is a collective effort. This section explores how individuals, businesses, and governments can contribute to a transition to sustainability, and highlights the importance of environmental education and awareness.

Individual Actions to Reduce the Ecological Footprint

Personal decisions and habits have a significant impact on the environment. While individual change cannot solve the problem alone, individual actions can contribute to collective change, influencing others to follow suit and generating demand for more sustainable products and practices. Here are some practical steps that each person can take to reduce their ecological footprint:

- **Reducing Consumption of Meat and Animal Products**: Meat production, especially beef, generates large amounts of greenhouse gases and consumes a large amount of water and land. Opting for a diet with less meat or adopting a plant-based diet can significantly reduce the ecological footprint. In addition, choosing local and seasonal products contributes to reducing CO_2 emissions associated with food transport.

- **Minimize the Use of Plastics and Single-Use Products**: Single-use plastics, such as bags, bottles, and wrappers, take hundreds of years to decompose and are responsible for the pollution of seas and soils. Adopting reusable alternatives, such as water bottles and cloth bags, and avoiding the use of single-use products are actions that each person can take to reduce their impact on the planet.

- **Energy Use from Renewable Sources and Energy Savings**: If possible, opting for renewable energy sources in the home, such as solar or wind, helps reduce dependence on fossil fuels. In addition, adopting energy-saving habits, such as turning off lights and appliances when not in use, installing energy-saving light bulbs, and optimizing the use of heating and air conditioning, are measures that reduce energy consumption and CO_2 emissions.

- **Sustainable Transportation**: Transportation is one of the main sources of greenhouse gas emissions. Choosing to walk, bike, use public transport or carpool with others reduces emissions and contributes to a cleaner urban environment. In case of using personal vehicles, considering electric or hybrid options also reduces the environmental impact.

Every little action counts. Adopting sustainable practices on a day-to-day basis is a way of taking individual responsibility, demonstrating that each person can contribute to the solution and encouraging others to do the same.

The Role of Business and Governments in Sustainability

To achieve meaningful change towards sustainability, it is essential that companies and governments adopt policies and practices that support environmental stewardship. The private and public sectors have the power to implement large-scale and structural changes that facilitate the transition to a more sustainable economy and society.

- **Public Policy and Environmental Legislation**: Governments play a critical role by creating policies and regulations that incentivize sustainable practices. Laws such as banning single-use plastics, regulating industrial emissions, and implementing carbon taxes are some of the measures that can reduce environmental impact. International commitments, such as the Paris Agreement, also set global targets to reduce emissions and limit global temperature rise.

- **Circular Economy and Waste Reduction**: The circular economy is a model that seeks to reduce waste and maximize the reuse of materials. Companies can adopt circular economy practices by designing durable and recyclable products, reducing material use, and minimizing waste throughout their supply chain. In addition, governments can promote this model through tax incentives and support for innovation in technology and sustainable production.

- **Investment in Renewable Energy and Corporate Sustainability**: Companies have an important responsibility in the transition to clean energy sources and in the adoption of sustainable practices. More and more corporations are committing to reducing their carbon footprint and using renewable energy in their operations. Leading companies in various sectors have set carbon neutrality goals, investing in green technologies and supporting carbon offset projects. These initiatives not only benefit the environment, but also contribute to a positive corporate image and respond to the growing consumer demand for sustainability.

Businesses and governments have the power to influence the structural change needed to address the environmental crisis. By implementing sustainable policies and adopting more responsible business models, they can lay the foundation for a more balanced and resilient future.

Educate and Create Environmental Awareness

Environmental education is an essential tool for building a society that values and protects the planet's natural balance. Environmental awareness fosters understanding of ecological issues and people's commitment to adopting sustainable practices and engaging in collective solutions.

- **Incorporating Environmental Education in Schools**: Including environmental education in school curricula is essential to train future generations in environmental challenges and solutions. Schools can teach students about climate change, biodiversity, ecosystems, and sustainability, and foster a critical and environmentally friendly mindset. This can include hands-on activities such as recycling projects, school gardens, and the use of renewable energy in schools.

- **Community and Media Awareness Campaigns**: Media and local community awareness campaigns are effective tools for informing the population about the importance of sustainability. Initiatives such as Earth Day, Earth Hour, and sustainability fairs create spaces for people to learn, share experiences, and be inspired to participate in activities that promote the conservation of the planet.

- **Training Programs for Leaders and Companies**: In addition to general education, it is important for business and government leaders to receive training in sustainability and environmental responsibility. Training programs in sustainable practices, circular economy and corporate social responsibility help decision-makers understand the importance of protecting the environment and adopt business strategies and public policies that favor sustainable development.

Education and environmental awareness are essential to building a culture of respect for the planet. By empowering people with knowledge and tools, a more engaged and proactive society is fostered in the fight against climate change and the protection of ecosystems.

Rebuilding the natural balance requires the combined effort of individuals, companies and governments, along with a solid environmental education that drives collective awareness and responsibility. From changes in daily habits to large-scale sustainability policies, every action counts in creating a sustainable future and restoring the harmonious relationship between humanity and nature.

KEY SECTORS FOR CHANGE

Renewable Energy and Energy Transition

The transition to renewable energies is one of the fundamental pillars in the fight against climate change. Clean energy is not only a sustainable alternative to fossil fuels, but it also represents an opportunity to create a safer, cheaper, and more efficient energy system. Below, the types of renewable energy, their advantages, the challenges of their adoption and the policies that are accelerating this transition are discussed in depth.

Introduction to Clean Energy and its Importance

Renewable energies are energy sources that are naturally regenerated and have a significantly lower environmental impact compared to fossil fuels. The main sources of renewable energy and their role in generating electricity without greenhouse gas emissions are highlighted below:

- **Solar Energy**: Solar energy converts sunlight into electricity through photovoltaic panels or heat through solar thermal systems. This source is abundant and available almost everywhere in the world, although it depends on direct sun exposure. Solar energy produces no emissions during use and has a very low environmental impact.

- **Wind Energy**: Uses the force of the wind to generate electricity through wind turbines. Wind energy is one of the fastest growing renewable sources in recent decades. As a clean energy source, it significantly reduces CO_2 emissions and can be installed both onshore and offshore.

- **Hydroelectric Energy**: It takes advantage of the energy of moving water, usually from rivers or reservoirs, to produce electricity. Although it is a source of clean energy, the construction of dams can have environmental impacts, such as the alteration of aquatic ecosystems and the displacement of communities. However, it is a stable and reliable source of energy in regions with water resources.

- **Geothermal energy**: Uses the Earth's heat to generate electricity or heat buildings. Geothermal energy is particularly effective in areas with high geothermal activity, such as Iceland and New Zealand. It is a continuous and stable source of energy, although its availability is geographically limited.

- **Biomass**: It consists of burning organic waste, such as wood, agricultural waste or plant waste, to produce energy. Biomass is considered renewable as long as the resources used are managed sustainably, as it releases CO_2 when burned. However, unlike fossil fuels, the CO_2 released can be reabsorbed by the growth of new plants.

Renewable energies are crucial to reduce the global carbon footprint and to mitigate climate change. By not relying on the burning of fossil fuels, these energy sources minimize greenhouse gas emissions and help reduce air pollution, benefiting both the environment and public health.

Advantages of Renewable Energies over Fossil Fuels

Renewable energies have a number of advantages compared to fossil fuels, both in environmental, economic and social terms:

- **Inexhaustible character**: Unlike oil, gas and coal, renewable energies are inexhaustible and are continuously regenerated. This makes renewable sources sustainable in the long term and secures energy supply without depleting natural resources.

- **Reduced Environmental Impact**: Renewables produce much lower carbon emissions than fossil fuels, making them critical options for combating climate change. In addition, renewables generate less air and water pollution, reducing negative effects on biodiversity and ecosystems.

- **Energy Independence**: Generating energy from renewable sources allows countries to reduce their dependence on imported fossil fuels, such as oil and gas. This increases energy security and protects economies from fluctuations in fossil fuel prices in international markets.

- **Job Creation**: The transition to renewable energy is generating employment opportunities in sectors such as manufacturing, installation and maintenance of clean energy infrastructure. According to the International Renewable Energy Agency (IRENA), the renewable energy sector employs millions of people globally and this figure is expected to continue to rise.

- **Economic Development and Accessibility**: Renewable energy can be especially beneficial in isolated or developing regions, where access to electricity is limited. Solar and wind systems, for example, can provide decentralized and affordable energy to rural communities, boosting economic development and improving quality of life.

These advantages position renewable energies as a viable and necessary option to ensure a sustainable and accessible energy future for all.

Challenges of the Energy Transition

Despite its benefits, the transition to a renewables-based energy system faces a number of technical, economic and logistical challenges:

- **Intermittency of some Renewable Sources**: Sources such as solar and wind depend on weather conditions and do not always produce energy continuously. This intermittency poses the challenge of ensuring a stable supply, especially in periods of low generation or high demand.

- **Need for Energy Storage Infrastructure**: To solve the problem of intermittency, it is essential to develop and improve energy storage technologies, such as high-capacity batteries. Lithium batteries, for example, are a promising solution, but their production depends on materials such as lithium and cobalt, the extraction of which can have environmental and social effects.

- **Reliance on Scarce and Critical Materials**: The production of renewable energy technologies, such as solar panels and wind turbines, requires specific materials that are not available in all regions. This can lead to dependency on certain producing countries and affect the availability of these materials as demand increases.

- **Start-Up Costs and Financing**: Although the cost of renewable energy has declined in recent years, upfront investment in infrastructure and technologies remains high. This is a challenge for many countries and communities that do not have sufficient financial resources to carry out this transition.

- **Just Transition and Community Resilience**: The transition to renewable energy can affect people working in fossil fuel industries. Just transition policies need to be implemented to help these workers adapt and ensure that the energy transition is inclusive and benefits all communities.

Technology investment and innovation in energy storage are essential to overcome these challenges and enable renewables to be effectively integrated into the global energy system.

Policies and Strategies to Accelerate the Transition

To drive the transition to renewable energy, it is necessary to implement public policies and strategies that promote their adoption and reduce dependence on fossil fuels. Some of the key policies and strategies in this process are outlined below:

- **Renewable Energy Subsidies and Incentives**: Many governments are implementing subsidies for renewable energy, such as tax breaks, investment subsidies, and feed-in tariffs that ensure favorable prices for renewable energy. These incentives help reduce investment costs and encourage the development of clean energy projects.

- **Emissions Regulation and Standard-Setting**: Regulating greenhouse gas emissions in sectors such as transportation, industry, and electricity generation is crucial to reducing the carbon footprint. Some nations are implementing emissions trading systems, carbon taxes, and CO_2 emission caps, which incentivize companies to reduce their emissions and adopt cleaner energy sources.

- **International Commitments and the Paris Agreement**: The Paris Agreement is a global pact in which countries commit to limit the increase in global temperature to less than 2°C, and, if possible, to 1.5°C. This agreement encourages nations to set ambitious emission reduction targets and develop action plans for the energy transition, including the development of renewable energies.

- **Smart Grid Infrastructure Development and Innovation**: Smart grids are electricity grids that optimize energy use through real-time control and communication. These grids allow for more efficient integration of renewable energy sources and improve the stability of the electricity system. The development of smart grids is essential to manage the intermittency of renewables and to ensure a reliable and efficient energy supply.

- **Clean Technology Research and Development Programs**: Investing in research and development is crucial to improving the efficiency and reducing costs of renewable technologies. Research programs in energy storage, marine energy, biofuels, and smart grids are contributing to technological advancement and the creation of more sustainable solutions for the energy system.

These policies and strategies are facilitating the transition to an energy system based on renewable sources and promoting a more sustainable, cleaner and safer energy future. Global cooperation and the commitment of governments, businesses and citizens are essential to accelerate this transition and achieve the climate goals needed to protect the planet.

Industry and Decarbonization

Industry is one of the sectors with the greatest responsibility for the generation of greenhouse gas emissions due to its energy consumption, production processes and waste generation. However, it also has enormous potential to reduce its environmental impact by adopting decarbonization practices, energy efficiency, and circular economy models. This section explores in depth the industry's role in global emissions, decarbonization strategies, and the policies and commitments that drive companies toward a more sustainable future.

The Industry's Role in Global Emissions

The manufacturing, transportation, and construction industries contribute to a large portion of global greenhouse gas emissions. The factors that explain this contribution are analyzed below:

- **Energy Consumption and Fossil Fuel Use**: Industry relies heavily on fossil fuels, such as coal, oil, and natural gas, for its production and transportation processes. This energy consumption is one of the main generators of CO_2, a greenhouse gas that contributes to climate change. The burning of fossil fuels to generate heat in sectors such as steel, cement and glass production accounts for a high percentage of industrial emissions.

- **Chemical Processes That Release Polluting Gases**: In addition to the combustion of fuels, many industrial processes generate polluting gases as byproducts. For example, cement production releases CO_2 during the calcination of limestone, and chemical manufacturing generates emissions of nitrous oxide and other greenhouse gases.

- **Impact of Industrial Transport**: The transport of goods and materials is another significant source of emissions, especially in global trade. Transporting goods through trucks, ships and planes, which also rely on fossil fuels, contributes to the carbon footprint of the industrial sector.

- **Responsibility and Potential for Change**: Despite its impact, the industry has great potential to reduce its emissions by adopting decarbonization practices and technological innovation. Companies across a variety of industries are beginning to take steps to mitigate their environmental impact, which is essential to achieving global emissions reduction targets.

Decarbonization of Industrial Processes

Industrial decarbonisation involves the reduction or elimination of carbon emissions in production processes, through the use of clean energy sources, the optimisation of processes and the minimisation of waste. Here are some examples of how companies are adopting decarbonization practices:

- **Switching Energy Sources**: One of the most effective ways to reduce industrial emissions is to replace fossil fuels with renewable energy sources. Many companies are installing solar panels and wind turbines to supply electricity to their factories, reducing their dependence on the power grid and lowering their carbon emissions.

- **Process Electrification**: In industries such as steelmaking, which have traditionally relied on fossil fuels to heat furnaces, process electrification is an emerging solution. For example, the use of electric furnaces instead of gas furnaces for steel production significantly reduces CO_2 emissions.

- **Carbon Capture and Storage (CCS):** Carbon capture and storage technology allows CO_2 emitted during industrial processes to be captured before it reaches the atmosphere, and then stored in underground geological formations. This technology is particularly useful in sectors that are difficult to decarbonize, such as cement and steel production.

- **Process Optimization and Waste Reduction**: The optimization of production processes, through the use of advanced technologies and best practices, allows companies to reduce energy use and reduce waste. Adopting more efficient production practices is critical to reducing the carbon footprint and improving the sustainability of the industry.

Energy Efficiency and Technology in Industry

Energy efficiency is a key strategy for reducing energy consumption in the industrial sector, and technology plays an essential role in its implementation. The adoption of advanced technologies, such as artificial intelligence (AI) and the Internet of Things (IoT), is transforming the way businesses manage their operations:

- **Process Optimization with AI and Big Data**: Artificial intelligence and data analysis allow companies to optimize their processes by monitoring energy use, material consumption, and equipment performance in real time. For example, AI systems can identify patterns of energy consumption and recommend adjustments to improve efficiency.

- **Internet of Things (IoT) for Intelligent Management**: IoT devices allow businesses to collect and analyze data on energy and resource usage in real-time. This data helps to identify areas for improvement and implement adjustments to reduce consumption and improve sustainability.

- **Automation and Robotics in Production**: Automating processes through the use of robots and AI-controlled machinery allows for improved accuracy and reduced material waste. This not only reduces environmental impact, but also increases productivity and lowers operating costs.

- **Energy Monitoring and Waste Management Systems**: Companies can implement monitoring systems to measure energy consumption and the volume of waste generated. These systems make it possible to establish goals for reducing consumption and recycling, promoting more efficient management of resources and a reduction in the carbon footprint.

Circular Economy and Waste Reduction

The circular economy is a model that seeks to reduce the use of materials and minimize waste generation through reuse and recycling. Unlike the traditional linear economy model, which is based on "produce, consume and dispose", the circular economy promotes a "reduce, reuse and recycle" approach:

- **Reuse and Recycling of Materials**: Companies can implement waste recycling and material reuse programs in their production processes. For example, in the textile industry, some companies are reusing waste fibers to manufacture new garments, thus reducing the demand for raw materials and the generation of waste.

- **Design for Durability and Repairability**: Adopting designs that allow for durability and easy repair of products is essential to reduce waste generation. In sectors such as electronics, companies are beginning to design modular and repairable products, which extend their useful life and minimize environmental impact.

- **Use of by-products as inputs**: The circular economy also involves using by-products of one process as inputs in another. In the food industry, for example, organic waste can be transformed into biogas for energy generation or compost to improve agricultural soils.

- **Reducing Packaging and Using Sustainable Materials**: Designing sustainable packaging, which minimizes the use of single-use plastics and favors biodegradable materials, is a common practice in companies looking to reduce their environmental impact and meet sustainability standards.

Corporate Commitments and Environmental Regulation

More and more companies are making commitments to sustainability and carbon neutrality, driven by both environmental regulations and consumer demand for responsible practices. These commitments and regulations play an important role in the transition to a more sustainable industry:

- **Carbon Neutrality and Emission Reduction Targets**: Many companies have adopted carbon neutrality commitments, setting goals to reduce their emissions and offset those they cannot avoid. Some are investing in reforestation and renewable energy projects to offset their emissions and achieve neutrality.

- **Emissions Regulation and Environmental Regulations**: Governments are implementing regulations that regulate industrial emissions and set limits on pollution. Carbon taxes, which tax companies according to their CO_2 emissions, are an effective tool to incentivize emissions reductions and promote the adoption of clean energy.

- **Sustainability and Corporate Social Responsibility Certifications**: Environmental certifications, such as ISO 14001 and eco-friendly product labels, allow companies to demonstrate their commitment to sustainability. These certifications, along with corporate social responsibility initiatives, help companies build a positive image and attract sustainability-conscious consumers.

- **Public Incentives and Sustainability Funds**: Many governments offer incentives, such as grants and tax deductions, to support companies in their transition to sustainability. These incentives facilitate investment in clean technologies and sustainable practices, accelerating the decarbonization of industry.

The combination of corporate commitments, government regulations, and incentive policies is driving companies to adopt more sustainable practices. This shift towards a decarbonised and responsible industry is essential to achieving climate goals and protecting the planet's balance.

Sustainable agriculture

Agriculture is an essential activity for human survival, but the intensification of unsustainable practices has led to significant negative impacts on the environment and climate. However, there are alternatives in sustainable agriculture that can mitigate these effects, preserve resources and promote more environmentally friendly food production. This section explores the impact of intensive agriculture, sustainable practices, agroecological and regenerative approaches, technological innovation, and government support for the transition to more environmentally friendly agriculture.

The Impact of Intensive Agriculture on Climate Change

Intensive agriculture, aimed at maximizing food production and especially focused on meat production, has had adverse consequences on the environment and the global climate. Some of the main impacts include:

- **Deforestation for Agricultural Land Expansion**: The growing demand for food, especially meat and animal products, has led to massive deforestation in regions such as the Amazon, where large areas of forest are cleared to create pastures for cattle and soybean cropping land (mostly used as animal feed). Deforestation not only reduces the planet's ability to absorb CO_2, but also causes massive biodiversity loss and alters natural water and nutrient cycles.

- **Methane and Greenhouse Gas Emissions**: The production of meat, especially beef, generates large amounts of methane, a greenhouse gas much more potent than CO_2 in terms of warming capacity. Methane is released mainly through the digestion of ruminant animals, as well as in manure management in intensive livestock systems.

- **Extensive Use of Fertilizers and Pesticides**: Intensive agriculture uses synthetic fertilizers and pesticides to maximize crop yields. These chemicals can release nitrogen oxides, another potent greenhouse gas, and also contaminate soils and water sources. Their continued use also deteriorates soil fertility and causes resistance in pests, generating the need to use increasing quantities.

- **Soil Degradation and Biodiversity Reduction**: The expansion of monocultures and the use of heavy machinery cause soil degradation and compaction, which decreases its capacity to store carbon and water. Biodiversity, which is crucial for the health of agricultural ecosystems, is also affected, as intensive agriculture reduces the habitats of native species and alters natural balances.

Sustainable Agriculture Practices to Mitigate Climate Change

To reduce the environmental impact of agriculture, a number of sustainable practices have been developed that help minimize emissions and protect natural resources:

- **Crop Rotation and Diversification**: Crop rotation involves alternating different types of crops on the same land to prevent soil depletion and reduce the need for pesticides. This method improves soil health, increases biodiversity, and reduces emissions associated with intensive agriculture.

- **Use of Organic Fertilizers**: Instead of synthetic fertilizers, organic fertilizers, such as compost and manure, offer a sustainable alternative. These not only reduce greenhouse gas emissions, but also improve soil structure and fertility in the long term.

- **Proper Agricultural Waste Management**: Responsible management of agricultural waste, such as crop residues and manure, is critical to reducing emissions of gases such as methane and nitrogen oxides. In some cases, agricultural waste can be used to produce biogas, a renewable energy source that reduces the use of fossil fuels.

- **Conservation Agriculture**: This approach includes techniques such as no-till and vegetation cover, which protect the soil from erosion, improve carbon sequestration, and maintain moisture. By reducing soil disturbance, conservation agriculture allows carbon to be retained in the soil, helping to mitigate climate change.

Agroecology and Regenerative Agriculture

Agroecology and regenerative agriculture are approaches that promote a harmonious relationship between agricultural activity and the natural environment, based on principles of sustainability and conservation:

- **Agroecology**: Agroecology seeks to integrate ecological principles into agricultural production, promoting production systems that maintain biodiversity and soil health. Instead of relying on external inputs, such as synthetic fertilizers and pesticides, agroecology uses natural techniques, such as biological pest control and crop rotation.

- **Regenerative Agriculture**: This approach goes beyond sustainability and focuses on regenerating degraded ecosystems. Regenerative agriculture promotes techniques such as planting trees in agricultural fields (agroforestry), the use of cover crops, and managed grazing, which improve soil fertility, increase its capacity to store carbon, and restore biodiversity.

- **Agroforestry Techniques**: Planting trees on agricultural land helps conserve soil moisture, provides shade, and protects crops from erosion. In addition, trees act as carbon sinks, reducing the amount of CO_2 in the atmosphere and improving the resilience of agricultural ecosystems.

- **Cover Crops and Pesticide Reduction**: Using cover crops, such as legumes or grasses, helps enrich the soil, prevent erosion, and reduce pesticide dependence, as these crops can compete with weeds and attract beneficial insects.

Technological Innovation in Agriculture

Technology is revolutionizing agriculture and enabling more efficient use of resources, facilitating more sustainable and environmentally friendly practices:

- **Drones and Aerial Monitoring**: Drones make it possible to monitor crops and obtain accurate data on soil condition, plant growth, and the presence of pests. This information helps farmers make informed decisions and apply inputs only where they are really needed, reducing the use of pesticides and fertilizers.

- **Precision Soil and Irrigation Sensors**: Soil sensors measure humidity, temperature, and other key factors that allow irrigation systems to be adjusted to provide only the amount of water needed. Precision irrigation helps reduce water consumption, which is critical in areas where water resources are limited.

- **Artificial Intelligence and Big Data in Farm Management**: Artificial intelligence and big data analytics allow farmers to predict weather patterns, optimize planting and harvesting, and manage resources more efficiently. These technological advances help increase productivity and reduce waste.

- **Precision Agriculture**: This approach uses advanced technology to track crop resources and needs in detail, allowing inputs (such as fertilizers and pesticides) to be applied accurately and in minimal quantities. This reduces environmental impact and increases efficiency in the use of resources.

Farmer Support and Incentive Policies

To foster a transition to sustainable agricultural practices, the support of governments and organizations is critical. There are a number of policies and programs that can help farmers in this process:

- **Subsidies and Incentives for Sustainable Agriculture**: Governments can offer subsidies and financial assistance to farmers who implement sustainable practices. These incentives may cover the cost of efficient irrigation technologies, the purchase of organic fertilizers, or the implementation of crop rotation systems.

- **Payment for Environmental Services (PES):** This program allows farmers to receive financial compensation for activities that contribute to environmental conservation, such as reforestation, soil restoration, and protection of water sources. This type of initiative helps to encourage the conservation and regeneration of natural resources.

- **Training and Technical Advice**: Organizations and governments can offer training and technical advice on sustainable practices, agricultural technology, and environmental management. This allows farmers to acquire the knowledge and skills needed to adopt sustainable practices and make more efficient use of resources.

- **Reforestation and Conservation Policies**: Government programs that incentivize reforestation on agricultural land and the conservation of natural areas near farmland help protect biodiversity and improve the soil's capacity to store carbon. These policies also support adaptation to climate change by reducing soil erosion and improving water retention.

Supporting farmers in the transition to sustainable agriculture is crucial to ensure that this industry can adapt to environmental challenges and continue to provide food responsibly and sustainably.

Path to a Sustainable Future: Collaboration

The challenge of mitigating climate change and securing a sustainable future requires active collaboration between the energy, industry, agriculture and other key actors. Only through efficient coordination and a shared vision can we achieve a low-carbon economy and preserve natural resources for future generations. This section discusses the interconnectedness of these sectors, highlights examples of collaborative projects, and emphasizes the need for collective commitment.

Interconnection between Sectors for an Efficient Transition

The transition to a sustainable economy is highly dependent on cooperation between sectors, as energy, industry and agriculture are closely interlinked and influence each other. Each sector brings a unique set of resources, technologies and knowledge that, combined, make it possible to reduce emissions and optimise the use of resources in a comprehensive way. Some examples of this interdependence and its benefits include:

- **Renewable Energy in Agriculture**: The adoption of renewable energy sources in the agricultural sector is critical to reducing dependence on fossil fuels in irrigation, production, and storage activities. Solar panels, for example, can be installed on agricultural land to supply electricity to irrigation systems, reduce costs and minimise CO_2 emissions associated with agricultural machinery and transport.

- **Use of Recycled Materials in Industry**: Industry has the ability to reuse materials from other sectors, which decreases the need to extract new raw materials and reduces waste. The construction sector, for example, can take advantage of recycled materials such as concrete and reclaimed steel, thus reducing its carbon footprint. This approach is not only more economical, but also promotes the circular economy.

- **Carbon Capture Technologies and Regenerative Agriculture**: The energy and industrial sectors can benefit from collaborating with regenerative agriculture to manage carbon emissions. Agricultural practices that increase soil carbon sequestration help offset emissions from other sectors, and at the same time improve soil quality, which increases the resilience of agricultural areas.

- **Grid Interconnection and Sustainable Energy in Industry**: Interconnecting renewable energy grids between companies and sectors allows industries to consume energy from clean sources, such as solar and wind. This integration reduces the demand for fossil fuels and allows for more efficient and sustainable production.

Cross-sector collaboration allows sustainable practices to be extended to all production areas and multiplies the positive impact of each initiative, creating a more integrated and resilient low-carbon economy.

Examples of Multisectoral Projects and Alliances

In recent years, numerous projects and partnerships have emerged between governments, businesses and non-governmental organizations that have demonstrated the potential of multi-sectoral collaboration to move towards a sustainable model. Some prominent examples include:

- **Alliance for the Decarbonization of Industry**: This initiative brings together companies from various industrial sectors with the aim of reducing emissions through clean technologies and best practices. Through the adoption of renewable energy and the development of energy efficiency solutions, the alliance seeks to set a global standard for the decarbonization of heavy industry, contributing to emission reduction targets in developed and developing countries.

- **Multi-Sector Reforestation Projects**: Reforestation projects, such as the Great Green Wall in Africa, are an example of collaboration between governments, international organizations, and local communities to restore ecosystems and combat desertification. This project, which covers multiple African countries, has been supported by different entities that work together to plant trees and native vegetation, which contributes to mitigating climate change and improving the quality of life of local communities.

- **Sustainable Agriculture and Biodiversity Initiative**: Organizations such as FAO and the United Nations Environment Programme have launched partnerships with governments and companies in the agricultural sector to implement sustainable agricultural practices that reduce environmental impact and promote biodiversity conservation. These alliances seek to optimize the use of resources, reduce water waste and protect local ecosystems, creating a more resilient and environmentally beneficial agriculture.

- **Alliances for the Circular Economy in the Textile Industry**: The textile industry is one of the most polluting, and many companies are collaborating to implement circular economy practices. Initiatives such as the "Fashion Pact", which involves large fashion brands, promote waste reduction, recycling of materials and the use of sustainable fibres, as well as a commitment to reduce carbon emissions throughout the supply chain.

- **International Agreements for Emission Reductions**: At the government level, multilateral agreements such as the Paris Agreement and the Glasgow Climate Pact are examples of collaboration between countries to establish emission reduction commitments and develop energy transition policies. These agreements foster global cooperation and serve as a basis for the creation of national policies and strategies aimed at sustainability.

Multi-sector projects and partnerships show how collaboration can generate comprehensive, large-scale solutions to today's environmental challenges, creating a positive effect that extends to diverse sectors and communities.

Towards a Sustainable Future: Vision and Collective Commitment

Achieving a sustainable future requires a shared vision in which all sectors, from government and industry to civil society, align their goals towards protecting the environment and reducing emissions. This collective commitment is essential to building a world where economic growth, social well-being and environmental health coexist in balance. Key elements to achieving this future include:

- **Environmental Education and Awareness**: Promoting environmental education at all levels of society is critical to increasing awareness of the importance of sustainability. Sustainability training should be present in schools, universities and work environments, so that both young people and professionals understand environmental challenges and engage in proactive solutions.

- **Innovation and Corporate Responsibility**: Companies have an essential role to play in the transition to a sustainable future, and must take responsibility for reducing their environmental impact and adopting more circular production and consumption models. Innovation in clean technologies and sustainable practices is key for companies to minimise their carbon footprint and promote environmentally friendly development.

- **Government Engagement and Supportive Policies**: Governments play a crucial role in implementing incentive policies for renewable energy, resource conservation, and decarbonization. Creating stricter environmental regulations, supporting clean energy, and financing sustainable projects are key to driving change and establishing a regulatory framework that promotes sustainability.

108

- **Fostering Innovation and Sustainable Technology**: Investment in research and development of sustainable technologies, such as carbon capture, clean energy, and precision agriculture, is necessary to create effective solutions to environmental problems. Cross-sector collaborations can empower the creation of technology that optimizes resource use and minimizes environmental impact.

- **Citizen Engagement and Sustainable Lifestyles**: Each individual has a responsibility to adopt sustainable practices in their daily lives. Reducing resource consumption, recycling and supporting sustainable products and companies are actions that contribute to sustainability. A change in citizens' lifestyles, driven by increased environmental awareness, is key to the success of global change.

- **Fostering Global and Regional Partnerships**: Collaboration at the international level, as well as regional partnerships between countries, allows for the sharing of knowledge, resources, and technologies to address common problems. International cooperation is vital to address challenges such as climate change and biodiversity loss, which transcend national borders.

The creation of a sustainable future is only possible if all sectors and social actors actively participate and align their goals and actions around a shared vision. By working together, we can build a low-carbon economy, preserve natural resources, and ensure a livable planet for future generations.

GOVERNMENTS AND ENVIRONMENTAL POLICY

Governments play a crucial role in creating a regulatory framework and implementing policies that promote sustainability and environmental protection. The transition to a low-carbon economy and climate change mitigation depend, to a large extent, on the ability of governments to establish laws, regulations, and international agreements that drive climate action at the global, regional, and local levels. This chapter explores in detail the role of governments in environmental policy, ranging from laws and regulations to international commitments and specific initiatives in Europe and Spain.

Environmental Laws and Regulations

Environmental legislation is a fundamental component for the protection of the planet, as it establishes guidelines and restrictions that regulate the use of natural resources and activities that impact the environment. Governments use these laws to address environmental challenges such as biodiversity loss, pollution, waste management, and water scarcity. Below are some key areas where environmental laws and regulations play an essential role in sustainability.

Protection of Biodiversity and Ecosystems

Biodiversity is vital for the health of ecosystems, as it ensures the balance of natural processes, from plant pollination to climate regulation. Laws focused on the protection of biodiversity and ecosystems include several actions:

- **Creation of Protected Areas**: Many countries have established national parks, nature reserves, and marine protected areas to preserve key habitats and prevent ecosystem degradation. These areas limit human activities that can harm the natural environment, such as deforestation, mining, and agricultural expansion.

- **Endangered Species Protection Laws**: In many countries, the protection of endangered species is regulated by specific laws that prohibit the hunting, capture, and trading of endangered animals. These laws typically include conservation programs to restore populations of vulnerable species, as well as the promotion of captive breeding of some species.

- **Regulation of Hunting and Fishing**: Hunting and fishing laws regulate the seasons and permitted quantities of catch, preventing overexploitation of species and allowing the regeneration of populations. These regulations help maintain biodiversity in aquatic and terrestrial ecosystems, and are accompanied by monitoring and control measures.

- **Protecting Key Ecosystems**: Governments often identify critical ecosystems, such as wetlands, mangroves, primary forests, and coral reefs, and enforce strict regulations for their protection. Not only are these ecosystems home to a wide variety of species, but they also perform important ecological functions, such as CO_2 absorption and protection against flooding and erosion.

Emissions and Pollution Regulation

Emissions regulation is essential to reduce greenhouse gases and improve air quality, which has a direct impact on climate change and public health. Laws in this area include:

- **Emissions Limits for Key Sectors**: Governments set limits on emissions of CO_2 and other pollutants for industries such as manufacturing, transportation, and energy. Companies must comply with these limits, which often tighten over time, to reduce their environmental impact.

- **Carbon Taxes**: Carbon taxes are an effective tool to incentivize businesses and consumers to reduce their consumption of fossil fuels. This tax taxes CO_2 emissions and typically directs the funds raised to sustainability projects, such as investment in renewable energy and reforestation.

- **Emissions Trading Systems**: Some countries and regions, such as the European Union, have implemented emissions trading systems in which companies can buy and sell permits to emit a limited amount of polluting gases. These systems establish an emissions cap and allow companies that manage to reduce their emissions below this limit to sell their permits to others, promoting efficiency and reducing the carbon footprint.

- **Air Quality Regulations**: To protect public health, many cities and countries have implemented regulations that limit concentrations of pollutants in the air, such as ozone, sulfur dioxide, and nitrogen dioxide. These regulations require controls in sectors such as transport and industry and are especially important in densely populated urban areas.

Waste Management and Recycling Laws

Waste management is essential to reduce the environmental impact of the accumulation of garbage and to avoid soil and water pollution. Waste management laws typically include:

- **Recycling Regulations**: Many countries have established recycling regulations that require businesses and citizens to separate waste and reuse materials. These laws facilitate the circular economy and help reduce the amount of waste that ends up in landfills, promoting the use of recycled materials in industry.

- **Ban on Single-Use Plastics**: Banning single-use plastics, such as bags, straws, and bottles, is an increasingly common measure to reduce plastic pollution. Governments are looking to replace these products with biodegradable and reusable alternatives, thereby decreasing the environmental impact of plastics on ecosystems.

- **Hazardous Waste Regulation**: Hazardous waste laws regulate the storage, transportation, and disposal of toxic materials, such as chemicals, medical waste, and e-waste. These regulations protect the environment and public health, requiring that hazardous waste be managed safely and responsibly.

- **Incentives for the Circular Economy**: In addition to recycling regulations, some governments promote the circular economy through tax incentives and subsidies for companies that implement sustainable production models. These policies support the reuse of materials and the reduction of waste, contributing to a more sustainable production and consumption model.

Regulations on Water Use and Protection of Water Resources

Water management is essential for sustainability and for meeting the needs of communities, agriculture and industry. Water resource protection laws include:

- **Regulation of Water Use in Specific Sectors**: In many countries, water use is regulated in intensive sectors such as agriculture and industry. These laws limit water extraction in periods of drought and establish rules for its responsible use, in order to prevent the depletion of reserves and protect aquatic ecosystems.

- **Protection of Water Sources and Recharge Zones**: Water source protection laws seek to preserve aquifers, lakes, rivers, and other important sources. In sensitive areas, such as aquifer recharge zones, restrictions are implemented to prevent pollution and ensure the quality of water for human consumption and biodiversity.

- **Wastewater Treatment Regulations**: Wastewater treatment legislation requires industries and urban areas to treat their water before returning it to water bodies. This prevents dangerous pollutants from reaching rivers and seas, and protects the health of communities and aquatic ecosystems.

- **Sustainable Watershed Management**: Many countries apply watershed management laws to protect areas that supply water resources. These laws promote sustainable land-use practices and watershed reforestation, which contributes to aquifer recharge and reduces flood risk.

Renewable Energy Grants and Incentives

To facilitate the transition to a low-carbon economy, governments implement subsidies and incentives for renewable energy. These policies include:

- **Solar and Wind Energy Grants**: Governments often offer subsidies to companies and individuals who install solar panels and wind turbines, making renewable energy more accessible and cost-effective. These subsidies help reduce the upfront cost of infrastructure and accelerate the adoption of clean energy sources.

- **Tax Incentives and Green Credits**: Tax incentives allow citizens and businesses to earn deductions or credits for investing in renewable energy projects. These policies make clean energy an economically attractive option and facilitate the energy transition.

- **Renewable Energy Research and Development Programs**: Many governments invest in research and development programs to improve renewable energy technologies. These programs fund innovation in energy storage, solar panel efficiency, and wind turbine upgrades, among others, and help reduce costs and increase the competitiveness of these energy sources.

- **Financing for Smart Grid Infrastructure**: Smart grids allow for more efficient management of energy generated by renewable sources and balance supply and demand. Governments finance smart grid projects that improve the integration of clean energy into the electricity system, reducing disruptions and facilitating the use of large-scale renewables.

Environmental legislation and regulations are essential to protect the environment and facilitate the transition to a sustainable economy. By setting clear standards in areas such as biodiversity, emissions, waste, water use, and renewable energy, governments create a framework for action that guides businesses and citizens towards more responsible and sustainable practices.

International Agreements and Global Commitments

Cooperation between nations is crucial to tackling environmental problems, especially climate change and biodiversity loss, which are borderless global challenges. Through international agreements, countries set common goals and share responsibilities to reduce emissions, protect biodiversity, and promote sustainable development. This section discusses some of the key agreements and commitments that guide climate and environmental action globally.

The Paris Agreement

The Paris Agreement is one of the most relevant climate agreements in the fight against climate change. Signed in 2015 during the United Nations Climate Change Conference (COP21) in Paris, this agreement established a global framework for the reduction of greenhouse gas emissions. Its main objectives include:

- **Limiting Global Temperature Rise**: The agreement sets out a commitment to limit global temperature rise to less than 2°C compared to pre-industrial levels, with the goal of not exceeding 1.5°C to avoid the most devastating impacts of climate change. This is critical to reducing the risk of extreme weather events, biodiversity loss, and other environmental effects.

- **Nationally Determined Contributions (NDCs):** Each country signing the Paris Agreement must regularly submit and update its NDCs, which are specific plans to reduce emissions and mitigate climate change. NDCs include emission reduction commitments, climate change adaptation plans and sustainability targets that must be reviewed and updated every five years, so that countries progressively increase their climate ambition.

- **Climate Finance for Developing Countries**: The Paris Agreement also promotes the mobilization of funds to help developing countries implement their NDCs and adapt to the effects of climate change. Developed countries have pledged $100 billion annually to finance climate action in developing countries, although this goal has yet to be fully met.

- **Transparency and Follow-up of Commitments**: The agreement establishes a system of transparency and accountability to ensure that countries report on their progress and meet their commitments. Progress reports make it possible to assess the achievement of the goals and increase pressure on governments to adopt more ambitious policies.

The Paris Agreement is a dynamic and revisable compact, allowing countries to increase their commitments over time based on needs and technological advances. It is a milestone in the history of climate cooperation and remains critical to achieving global sustainability goals.

Kyoto Protocol and its Historical Relevance

The Kyoto Protocol, signed in 1997 and in force since 2005, was the first international treaty to set binding emission reduction targets for developed countries. Although it was replaced by the Paris Agreement in 2015, the Kyoto Protocol marked a milestone in global climate policy:

- **Emission Reduction Targets for Developed Countries**: The protocol obligated developed countries to reduce their greenhouse gas emissions by an average of 5% compared to 1990 levels, during the period from 2008 to 2012. These binding targets set a precedent for future climate negotiations.

- **Creation of Carbon Market Mechanisms**: The Kyoto Protocol introduced the first carbon market mechanisms, such as the Clean Development Mechanism (CDM) and emissions trading. These mechanisms allowed countries and companies with emission reduction targets to offset their emissions by investing in carbon reduction projects in other countries, which fostered international cooperation and innovation in clean technologies.

- **Second Commitment Period**: The Kyoto Protocol was extended in 2012 by the Doha Amendment, which established a second commitment period until 2020. However, the protocol had limitations, as it did not include some of the largest emitters of greenhouse gases, such as the United States and China, which led to the need for a more inclusive agreement, such as the Paris Agreement.

The Kyoto Protocol pioneered the creation of a legal framework for emissions reductions and the implementation of market mechanisms for climate mitigation, laying the groundwork for current climate commitments.

Glasgow Climate Pact

The Glasgow Climate Pact was an agreement adopted during COP26 in 2021. This summit marked a milestone in global climate action, reinforcing the commitments of the Paris Agreement and underlining the urgency of immediate action to limit global warming to 1.5°C. Highlights of the pact include:

- **Reducing Coal Use and Eliminating Fossil Fuel Subsidies**: The pact calls for phasing out the use of coal, which is one of the largest emitters of CO_2, and eliminating inefficient fossil fuel subsidies. This commitment marks an important step towards the decarbonization of the global economy.

- **Increasing Climate Finance**: One of the objectives of the pact is to increase financing to help developing countries adapt to the effects of climate change. Goals were also set to improve the resilience of vulnerable communities, fostering climate justice and equity in climate action.

- **Commitment to Update NDCs by 2022**: Unlike the five-year deadline of the Paris Agreement, the Glasgow pact urged countries to update their NDCs in 2022 to reflect greater climate ambition, given the urgency to reduce emissions by 2030. This reinforces the importance of constant review and adaptation of climate commitments.

- **Protecting Ecosystems and Reducing Deforestation**: The pact also underscores the importance of natural ecosystems, promoting the protection of forests and reducing deforestation as essential measures to combat climate change and preserve biodiversity.

The Glasgow Climate Pact is a reminder of the urgency of action and a renewed commitment to achieving the Paris goals, focusing on rapid emissions reductions and financing for climate adaptation.

Agreements for the Conservation of Biodiversity

The conservation of biodiversity is a fundamental pillar of environmental sustainability. In addition to climate agreements, there are international conventions that seek to protect biodiversity and promote the sustainable use of natural resources:

- **Convention on Biological Diversity (CBD):** This convention, adopted at the Rio Earth Summit in 1992, aims at the conservation of biodiversity, the sustainable use of its components, and the fair and equitable sharing of benefits derived from genetic resources. Through action plans, countries implement conservation and sustainability measures, promoting the recovery of endangered species and the creation of protected areas.

- **Ramsar Convention on Wetlands**: Signed in 1971, the Ramsar Convention aims to protect wetlands of international importance, which are key ecosystems for biodiversity, water regulation and climate change mitigation. The signatory countries undertake to designate Ramsar sites and to implement measures for the protection and sustainable use of these ecosystems.

126

- **Convention on International Trade in Endangered Species of Wild Fauna and Flora (CITES):** This convention regulates trade in endangered species of fauna and flora to prevent their unsustainable exploitation. CITES establishes a permitting system for international trade in protected species, promoting the conservation of endangered species and the protection of ecosystems.

- **Agreement on the Conservation of African-Eurasian Migratory Waterbirds (AEWA):** This agreement, adopted in 1995, seeks to protect migratory waterbirds and their habitats in Europe, Africa, and Asia. Through conservation, monitoring and awareness-raising actions, AEWA promotes international collaboration to protect the migratory routes and habitats of these species.

These international conventions are fundamental for the protection of biodiversity and the balance of ecosystems, and establish a framework for global cooperation to address the biodiversity crisis.

UN Sustainable Development Goals (SDGs)

The Sustainable Development Goals (SDGs), established in 2015 by the UN General Assembly, are a roadmap for sustainable development at the global level. The SDGs include 17 goals that cover different aspects of environmental, social and economic sustainability, and some are directly related to climate action and environmental protection:

- **SDG 13: Climate Action**: This goal seeks to adopt urgent measures to combat climate change and its effects, aligning with the commitments of the Paris Agreement. It includes specific targets to strengthen resilience and adaptation to climate disasters and to integrate climate policies into national plans.

- **SDG 14: Life Below Water**: This goal promotes the conservation and sustainable use of the oceans, seas and marine resources. Goals include reducing marine pollution, protecting coastal ecosystems, and restoring fish stocks, which are critical to maintaining the balance of aquatic ecosystems.

- **SDG 15: Life on Land**: This goal focuses on the protection, restoration and sustainable use of terrestrial ecosystems, such as forests, mountains and wetlands. The goals include halting deforestation, restoring degraded ecosystems, and protecting biodiversity, essential elements for sustainable development.

- **SDG 6: Clean Water and Sanitation**: This goal promotes universal access to safe drinking water and sanitation, and sets targets to improve water quality, reduce pollution, and protect aquatic ecosystems. Sustainable water management is crucial for development and for tackling climate change.

- **SDG 7: Affordable and Clean Energy**: This goal promotes access to affordable, safe, sustainable and modern energy. Its focus on clean, renewable energy is key to reducing greenhouse gas emissions and advancing the energy transition.

The SDGs provide a comprehensive vision of sustainability and guide the actions of governments, companies and organizations towards balanced development that protects the environment and improves the quality of life.

European Initiatives for Sustainability

The European Union (EU) has set out a set of ambitious policies and initiatives to lead the transition to a sustainable future. With the aim of being the first climate-neutral continent by 2050, the EU seeks not only to reduce its own greenhouse gas emissions, but also to establish a model that inspires other regions of the world. This section analyses major European initiatives covering various sectors, from energy and the economy to biodiversity and sustainable project financing.

European Green Deal

The European Green Deal is a comprehensive plan that guides the transformation of the EU economy to make it more sustainable. Adopted in 2019, this pact focuses on achieving carbon neutrality by 2050 and reducing greenhouse gas emissions by at least 55% by 2030 compared to 1990 levels. Key areas of the Green Deal include:

- **Energy Transition**: The Green Deal drives the shift towards clean and renewable energies, reducing dependence on fossil fuels. To this end, the EU encourages the uptake of clean technologies, such as solar and wind power, and supports investment in energy storage infrastructure to ensure a stable supply.

- **Sustainable Mobility**: The sustainable mobility strategy promotes low-emission transport by promoting electric vehicles, developing charging infrastructures and improving public transport. In addition, it encourages the shift towards rail and maritime transport systems, which are less polluting compared to air and land transport.

- **Sustainable Agriculture and the Farm to Fork Strategy**: The Green Deal sets out the Farm to Fork strategy, which promotes sustainability in agriculture and food production. This strategy seeks to reduce the use of pesticides and fertilizers, promote organic production and ensure the sustainability of the food chain.

- **Ecosystem and Biodiversity Restoration**: The EU is also committed to restoring and protecting ecosystems and biodiversity, promoting reforestation, protecting marine areas and reducing soil and water pollution. Biodiversity conservation is a priority to mitigate climate change and improve the resilience of ecosystems.

The European Green Deal represents an ambitious framework that not only sets climate targets, but also integrates sustainability into all economic sectors, from energy to agriculture and transport, fostering a green and inclusive economy.

EU Taxonomy Regulation

The EU Taxonomy Regulation is a pioneering tool that defines which economic activities can be considered environmentally sustainable. This classification system provides clarity for investors, companies and governments, making it easier to finance projects that contribute to sustainability. Key elements of the taxonomy include:

- **Sustainability Criteria**: The regulation establishes clear criteria for classifying sustainable activities in sectors such as energy, transport, construction and waste management. Activities are evaluated based on their contribution to six environmental objectives, including climate change mitigation and adaptation, the circular economy and the protection of biodiversity.

- **Transparency for Investors and Companies**: The taxonomy provides investors with a tool to identify investment opportunities in projects and companies that meet sustainability criteria. This increases transparency in the financial market and facilitates the flow of capital into green activities.

- **Facilitating Sustainable Finance**: The taxonomy is a foundation for green bonds and other sustainable financial instruments, which allow investors to channel funds into projects that meet high environmental standards. This framework regulates the issuance of sustainable financial products, aligning investments with the EU's climate objectives.

The Taxonomy Regulation contributes to the transformation of the European economy by directing investment towards sustainable activities and fostering the transition to a carbon-neutral economy.

Circular Economy Strategy

The Circular Economy Strategy is a key initiative to reduce the exploitation of natural resources and minimise waste generation. This approach promotes the transition from a linear economic model ("produce, use, and dispose") to a circular model based on reduction, reuse, and recycling. Some key components of this strategy include:

- **Ecodesign and Sustainable Production**: The strategy promotes the design of more durable, repairable and recyclable products, encouraging production practices that minimize the use of resources and the generation of waste. This includes the regulation of electronics, textiles, and packaging, which must meet sustainability standards.

- **Sustainable Waste Management**: The circular economy encourages the separation and recycling of waste to reduce the amount of materials that end up in landfills. The EU seeks to increase recycling rates, in particular of plastics and hazardous materials, and reduce the use of single-use plastics in products and packaging.

- **Promoting Responsible Consumption**: The strategy also encourages a change in consumer behavior, promoting product repair, reuse, and responsible consumption. Initiatives such as return and deposit systems for packaging and the promotion of the purchase of recycled products are examples of how the circular economy seeks to involve consumers.

- **Innovation and New Business Models**: The circular economy drives innovation and the creation of business models based on rental, repair and recycling. This encourages the emergence of companies that promote reuse and recycling services, contributing to sustainability and the generation of green employment.

The circular economy strategy seeks to create a more efficient and sustainable economy, which reduces the exploitation of resources and minimizes the environmental impact of production and consumption.

LIFE Program

The LIFE Programme is the EU's main funding instrument for environmental conservation, climate action and sustainability projects. Since its creation in 1992, LIFE has funded thousands of projects across Europe, promoting innovation and the implementation of sustainable solutions. The main components of the program include:

- **Nature and Biodiversity Conservation**: LIFE finances conservation projects that protect and restore endangered ecosystems and species. These projects include habitat restoration, reforestation and the protection of natural areas, contributing to the preservation of biodiversity in Europe.

- **Climate Action and Climate Change Mitigation**: LIFE supports projects that seek to reduce greenhouse gas emissions through the implementation of renewable energies, the improvement of energy efficiency and the promotion of sustainable practices in sectors such as agriculture and industry.

- **Circular Economy and Resource Efficiency**: LIFE finances projects that promote the circular economy, promoting waste reduction and the reuse of materials. This includes initiatives to improve waste management and promote recycling in industrial and urban sectors.

- **Support for Citizen Participation and Environmental Awareness**: The program also funds projects that promote environmental education and awareness, involving communities in environmental protection. This fosters greater awareness of the importance of sustainability and climate action.

The LIFE Programme is a fundamental pillar in the financing of environmental initiatives in Europe, contributing to innovation and the implementation of effective solutions to face environmental challenges.

European Climate Law

The European Climate Law, passed in 2021, makes the EU's goal of being climate neutral by 2050 a legal obligation. This law sets binding emission reduction targets and requires Member States to report regularly on their progress. Key aspects of the European Climate Law include:

- **Long-Term Emission Reduction Targets**: The law states that the EU must reduce its emissions by at least 55% by 2030 and achieve climate neutrality by 2050. These targets are binding and reflect the EU's commitment to the Paris Agreement and the fight against climate change.

- **Adaptation to Climate Change**: The law not only focuses on mitigation, but also on adaptation, establishing that member states must implement measures to improve the resilience of their economies and communities to the effects of climate change. This includes plans to protect water resources, infrastructure, and public health.

- **Review and Monitoring of Progress**: The law requires Member States to report regularly on their progress in reducing emissions and implementing sustainability policies. The European Commission assesses these reports and may recommend additional measures to ensure that the targets are met.

- **Inclusion of Society and Economic Agents**: The European Climate Law seeks to involve civil society and the private sector in the transition to sustainability. This includes public consultations and the promotion of initiatives that encourage the participation of companies and citizens in the fight against climate change.

The European Climate Law provides a strong legal basis for climate action in Europe and reinforces Member States' commitment to a climate-neutral future.

These European initiatives show the EU's leadership in implementing ambitious environmental policies and its commitment to long-term sustainability. Through comprehensive and binding policies such as the Green Deal, the circular economy, the LIFE programme and the European Climate Law, the EU not only establishes a framework for environmental protection, but also drives an economic transformation that promotes innovation, green job creation and a climate-resilient economy.

Environmental Policies in Spain

Spain, in line with the European Union's climate and environmental objectives, has implemented various policies and strategies to reduce its environmental impact and adapt to the effects of climate change. These policies address everything from the energy transition to the protection of biodiversity, the circular economy and adaptation to extreme events. Below are some of the main environmental policies in the country.

Climate Change and Energy Transition Law

Approved in 2021, the Climate Change and Energy Transition Law is the first regulation in Spain to establish emission reduction targets and promote the use of renewable energies to facilitate the transition to a low-carbon economy. Its highlights include:

- **Climate Neutrality Goal for 2050**: The law sets a goal of climate neutrality by 2050, which implies that greenhouse gas emissions must be reduced to the minimum possible and the rest must be offset. This objective is essential to align Spain with European commitments and the Paris Agreement.

- **Emissions Reduction by 2030**: An emissions reduction target of 23% by 2030 is set compared to 1990 levels. This commitment involves a significant effort in sectors such as energy, transport and industry, with policies that promote energy efficiency and the adoption of clean technologies.

- **Promotion of Renewable Energies**: The law promotes the expansion of renewable energies and establishes that at least 42% of the energy consumed in the country must come from renewable sources by 2030. In addition, investment in energy storage infrastructures and smart grids that facilitate the integration of these sources into the electricity system is encouraged.

- **Electrification of Transport and Sustainable Mobility**: The law establishes a ban on the sale of vehicles that emit CO_2 by 2040 and promotes the development of charging infrastructures for electric vehicles. In addition, sustainable mobility projects are encouraged and the improvement of public transport is promoted to reduce dependence on fossil fuels.

- **Protection of Ecosystems and Biodiversity**: The regulations also include measures for the protection and restoration of ecosystems, recognizing the importance of natural resources in mitigating climate change. The restoration of degraded areas and the creation of ecological corridors are part of this commitment.

The Climate Change and Energy Transition Law is one of the most important policies to guide Spain towards a more sustainable future and reduce its carbon footprint.

National Integrated Energy and Climate Plan (PNIEC)

The National Integrated Energy and Climate Plan (PNIEC) is a roadmap that sets out the actions that Spain will implement between 2021 and 2030 to meet its emission reduction commitments. This plan focuses on five priority areas:

- **Expansion of Renewable Energies**: The PNIEC establishes that 74% of electricity must be generated from renewable sources by 2030. This implies a significant increase in the installed capacity of solar, wind and other renewable sources, with investments in new infrastructure and technologies.

- **Improving Energy Efficiency**: The plan includes measures to reduce energy consumption by 39.5% by implementing energy efficiency technologies in buildings, transport and industry. Improving efficiency is key to reducing energy demand and decreasing emissions.

- **Reducing the Use of Fossil Fuels in Transport**: The PNIEC encourages the electrification of transport and the transition to zero-emission vehicles. To this end, incentives are established for the purchase of electric vehicles and the development of charging infrastructures is promoted, especially in urban and rural areas.

- **Research and Innovation in Clean Technologies**: The plan supports the research and development of clean technologies and energy storage solutions that facilitate the integration of renewables. This includes investment in energy storage systems and the digitalization of electricity grids.

- **Just Transition for Coal-Dependent Regions**: The PNIEC includes a just transition commitment to support communities and workers affected by the closure of coal mines and thermal power plants. This support involves the creation of jobs in sustainable sectors and the development of economic projects in regions in transition.

The PNIEC establishes ambitious goals that will allow Spain to advance in its energy transition and contribute to the reduction of emissions at the European level.

Circular Economy Strategy in Spain

The Spanish Circular Economy Strategy aims to transform the production and consumption model to reduce the use of materials, promote reuse and recycling and minimise waste generation. Key components of this strategy include:

- **Waste Reduction and Recycling**: The strategy sets specific waste reduction targets for 2030, with an emphasis on eliminating single-use plastics and improving recycling infrastructure in sectors such as construction, textiles and packaging.

- **Promotion of Reuse and Ecodesign**: Encourages the design of products that are more durable, repairable and recyclable, facilitating their reuse and reducing waste. Eco-design is especially applied in electronic products, textiles and packaging.

- **Fostering the Circular Economy in Industry and Construction**: The strategy promotes the use of recycled materials in industry and construction, as well as the creation of new business models based on the circular economy. This includes tax incentives and support for research into recycling technologies.

- **Citizen Participation and Awareness**: The strategy also seeks to raise awareness among the population about the importance of responsible consumption and the circular economy. Through education and awareness campaigns, the participation of citizens in the reduction of waste and the separation of materials is encouraged.

The Circular Economy Strategy contributes to reducing the environmental impact of consumption in Spain and to building a more sustainable and efficient economic model.

Natura 2000 Network

Spain is home to one of the largest areas of the Natura 2000 Network, a set of protected areas that covers more than 27% of the national territory. This network aims to conserve biodiversity and protect habitats and endangered species of flora and fauna. Key components of the Natura 2000 Network in Spain include:

- **Protection of Natural Habitats and Threatened Species**: The network protects specific habitats, such as forests, wetlands, mountains, and coastal areas, ensuring the survival of endangered species. Designated areas are subject to conservation and management measures that limit activities that may degrade the environment.

- **Regulation of Activities in Protected Areas**: The Natura 2000 Network regulates human activities in these areas, including agriculture, tourism and industry, to ensure that ecosystems are not damaged. The network promotes sustainable practices and respects the ecological values of each area.

- **Fostering Biodiversity and Healthy Ecosystems**: The network is a key tool for species recovery and restoration of degraded ecosystems. Conservation activities in these areas include reforestation, invasive species control, and habitat restoration to enhance biodiversity.

- **International cooperation**: The Natura 2000 Network is part of the EU's conservation strategy and is aligned with the Convention on Biological Diversity. Cooperation between EU countries facilitates the protection of migratory routes and the conservation of transboundary species.

The Natura 2000 Network in Spain is essential to preserve biodiversity and guarantee the protection of natural resources in the country.

Climate Change Adaptation Plan

Spain is one of the countries in Europe most vulnerable to the effects of climate change, such as droughts, desertification and rising temperatures. The National Plan for Adaptation to Climate Change aims to strengthen the country's resilience to these impacts and protect vulnerable communities. Key components of the plan include:

- **Measures for Water Management and Agriculture**: The plan promotes sustainable water use and adapts agriculture to changing climatic conditions. This includes implementing efficient irrigation technologies, managing watersheds, and diversifying crops that are more resistant to drought.

- **Public Health Protection**: The plan addresses health risks associated with climate change, such as increased respiratory and cardiovascular diseases due to heat waves and the spread of vector-borne diseases. Strategies are developed to improve the response capacity of health services to these risks.

- **Strengthening Infrastructure**: Transportation, energy, and building infrastructures are being adapted to withstand extreme weather conditions, such as heat waves, floods, and storms. This includes retrofitting buildings to improve energy efficiency and building more resilient infrastructure.

- **Protection of Vulnerable Ecosystems**: Measures are implemented to protect and restore ecosystems vulnerable to climate change, such as coastal areas, wetlands, and mountains. The adaptation of these ecosystems is crucial to preserve their biodiversity and their capacity to store carbon.

The Climate Change Adaptation Plan allows Spain to prepare for the effects of climate change, protecting the economy, biodiversity and the well-being of the population.

These policies reflect Spain's commitment to sustainability and environmental protection. From the transition to a low-carbon economy to biodiversity conservation and adaptation to climate impacts, Spain implements ambitious policies that not only contribute to European goals, but also improve the quality of life and resilience of its communities in the face of current and future environmental challenges.

International and Local Collaboration

Cooperation is essential to address global environmental challenges, which affect all regions of the planet without distinction of borders. Collaboration between countries, local governments and citizens plays a fundamental role in the implementation of effective policies that mitigate the effects of climate change, protect biodiversity and promote sustainable development. This section details the different forms of collaboration that are essential to address these environmental challenges.

Cooperation between Countries

International collaboration is crucial to share knowledge, technologies and resources that enable the creation of effective environmental policies. In the field of environmental policy, countries cooperate to coordinate their efforts within the framework of global agreements and strategic alliances:

- **Technology and Knowledge Transfer**: Collaboration allows countries with the greatest technological advances to share knowledge and clean technologies with developing nations. These technologies can include renewable energy, waste management techniques, and sustainable agriculture systems, which help reduce emissions and improve resource efficiency.

- **Clean Energy Research and Development**: International cooperation facilitates the joint creation and financing of research projects to develop new clean technologies, such as energy storage systems, carbon capture, and alternative energy sources. These joint efforts allow countries to advance innovation and find sustainable solutions to their energy needs.

- **Strategic Partnerships for the Conservation of Global Ecosystems**: Ecosystems such as the Amazon, oceans, and polar regions have a global impact. For this reason, several countries collaborate in conservation and monitoring programs for these environments, to protect their biodiversity and their role in climate regulation. Initiatives such as the Ocean Conservation Partnership enable coastal countries and international organizations to work together to protect marine ecosystems.

- **Setting Global Environmental Norms and Standards**: Through agreements such as the Kyoto Protocol, the Paris Agreement, and the Convention on Biological Diversity, countries establish common goals for reducing emissions, protecting biodiversity, and conserving natural resources. These global agreements make it possible to create shared standards and guidelines, ensuring that all countries take responsibility for protecting the environment.

Cooperation between countries strengthens global responsiveness to environmental challenges and enables nations to work together to achieve sustainability goals.

Collaboration between Regional and Local Governments

Regional and local governments play a key role in implementing environmental policies and adapting them to the specific needs of their communities. This collaboration allows policies to be implemented more effectively and to achieve a direct impact on the quality of life of citizens:

- **Sustainable Waste Management and Recycling**: Local governments are responsible for waste management in their communities. Collaboration between municipal and regional governments facilitates the creation of efficient recycling systems, the implementation of waste reduction campaigns, and the adoption of circular economy practices, such as composting and reusing materials.

- **Promoting Sustainable Mobility and Public Transport**: Local governments can implement sustainable transport policies by improving public transport systems, building bike lanes, and promoting electric vehicles. Collaboration between cities and regions allows for best practices to be shared and sustainable solutions to be implemented adapted to the characteristics of each area.

- **Protection of Natural Areas and Urban Green Zones**: Local governments play a fundamental role in the creation and protection of parks, reserves and green areas. These spaces not only contribute to biodiversity, but also offer benefits to the health and well-being of the population. Collaboration between different levels of government facilitates the financing and management of these areas.

- **Resilience to Natural Disasters and Climate Change Adaptation**: Local governments are the first to respond to the effects of extreme weather events, such as floods, heat waves, and droughts. Collaboration with regional and national governments allows for the implementation of resilience and adaptation programs that protect vulnerable communities and strengthen local infrastructure.

Collaboration between different levels of government ensures that environmental policies are implemented efficiently and that they are tailored to the needs and characteristics of each community.

Promotion of Citizen Participation and Transparency

Citizen participation is key for environmental policies to respond to the needs and expectations of the population, and to ensure the effectiveness and legitimacy of government decisions. Governments can promote transparency and participation through:

- **Public Consultations and Spaces for Participation**: Governments organize public consultations and forums in which citizens can express their opinions on environmental policies and propose solutions. These spaces allow communities to participate in decision-making, promoting an inclusive and participatory approach to policy-making.

- **Environmental Education and Awareness**: Awareness and education are essential to promote responsible environmental behavior in the population. Through environmental education campaigns in schools, the media and social networks, governments promote awareness of the importance of sustainability, the circular economy and waste reduction.

- **Transparency in Environmental Management**: Transparency is essential to ensure that environmental policies are responsible and reliable. Governments can provide access to environmental information through web portals, public reports, and open data on topics such as air quality, energy consumption, and progress in reducing emissions.

- **Volunteer Programs and Participation in Conservation Projects**: Volunteer programs in protected areas, beach and river cleanups, and reforestation activities are some of the ways in which citizens can become actively involved in environmental protection. These programs encourage the commitment of the population and strengthen the relationship between the community and the environment.

Citizen participation and transparency not only strengthen the legitimacy of environmental policies, but also contribute to creating a society that is more aware of and committed to sustainability.

Assistance and Financing for Developing Countries

Developing countries face significant challenges in implementing environmental policies due to economic and technological constraints. Developed countries have a responsibility to provide support through climate assistance and financing, which facilitate the adoption of climate change sustainability and adaptation policies:

- **Climate Finance for Mitigation and Adaptation**: International funds, such as the Green Climate Fund, provide resources for developing countries to reduce their greenhouse gas emissions and adapt to the effects of climate change. These funds finance renewable energy projects, reforestation, climate-resilient infrastructure, and other sustainable initiatives.

- **Technology and Knowledge Transfer**: Technology transfer is essential to help developing countries implement sustainable solutions. International collaboration facilitates access to advanced technologies, such as renewable energy, water management techniques and sustainable agriculture systems, which enable these countries to reduce their emissions and improve their resilience.

- **Training and Technical Assistance**: Cooperation includes training local professionals in areas such as waste management, energy efficiency and biodiversity conservation. This technical assistance helps build capacity in developing countries and strengthens their ability to implement effective environmental policies.

- **Green and Sustainable Infrastructure Development**: Developed countries can provide financing and technical support for the construction of sustainable infrastructure in developing countries, such as public transport systems, smart grids, and water treatment plants. These infrastructures enable a transition to a more sustainable and low-carbon economy.

- **Supporting the Adaptation of Vulnerable Communities**: Communities most vulnerable to climate change, such as coastal areas, arid regions, and areas of high biodiversity, require additional support to adapt to climate change. International cooperation facilitates the implementation of adaptation projects that protect livelihoods, biodiversity, and infrastructure in these communities.

Climate assistance and financing are essential for developing countries to meet their environmental commitments and address the effects of climate change effectively.

Collaboration between different levels of government and international cooperation are essential to address global and local environmental challenges. From technology transfer and climate finance for developing countries to citizen participation and regional and local policy coordination, collaboration enables a more effective and equitable response to the climate and environmental crisis. These partnerships strengthen the resilience of communities, improve quality of life, and promote a sustainable and equitable future for all.

SUSTAINABILITY IN DAILY LIFE

Sustainability in daily life is one of the most effective ways we can all contribute to the protection of the planet. Adopting sustainable practices in our daily routine not only helps reduce our carbon footprint, but also promotes a culture of respect for the environment. This chapter explores different areas where each person can take concrete actions to live more sustainably.

Sustainable Mobility

Transport is one of the sectors that contributes the most to greenhouse gas emissions and air pollution. Adopting sustainable mobility practices not only helps reduce emissions and protect the environment, but also improves the quality of life in our cities. There are different sustainable mobility options that can be adapted to the needs of each person and that offer a more responsible way of getting around. Some of the top sustainable mobility options are explored below.

Use of Public Transport

Public transport is one of the most effective alternatives to reduce the environmental impact of transport. Opting for buses, trains, metros and trams instead of private vehicles reduces CO_2 emissions and contributes to more sustainable and accessible mobility:

- **Reduced Emissions per Person**: Public transportation allows more people to be transported in fewer vehicles, which significantly decreases emissions per passenger. A bus or train carrying dozens of people emits less CO_2 per person compared to the same number of people commuting in individual cars.

- **Reducing Congestion in Cities**: By reducing the number of cars on the road, public transportation helps decrease urban congestion, improving traffic flow and reducing the time vehicles stay on the road. This not only benefits the environment, but also improves the quality of life of citizens.

- **Energy Savings and Reduction of Dependence on Fossil Fuels**: Public transportation systems can run on electric power and renewable sources, decreasing dependence on fossil fuels. More and more cities are deploying electric buses and trains powered by green energy to make public transport even more sustainable.

- **Accessibility and Economic Benefits**: Public transport is a more accessible and economical option for many people, allowing citizens of different socioeconomic levels to move around without the need for a private vehicle. In addition, as it is a public service, it contributes to more equitable and fair mobility.

Public transport represents a sustainable mobility option, and strengthening it through investments in infrastructure and quality of service is key to promoting its use.

Cycling and Walking as Sustainable Options

Cycling and walking are completely emission-free means of transport and offer multiple benefits for both the environment and people's health. These options are especially practical over short distances and in urban environments.

- **Zero-Emission Mobility**: Cycling and walking are forms of mobility that do not generate CO_2 emissions, making them the cleanest options from an environmental point of view. They reduce air pollution and help mitigate the effects of climate change.

- **Health Benefits**: Biking and walking improve physical and mental fitness, and help reduce the risk of cardiovascular disease, obesity, and stress. By integrating these practices into their daily routine, people can improve their well-being and reduce sedentary lifestyles.

- **Infrastructure for the Promotion of Cycling**: Many cities are investing in the creation of bike lanes, bike lanes, and bike sharing systems, which facilitates the use of bicycles as a safe and efficient means of transportation. Shared bicycles are an economical and accessible option that allows more people to opt for this means of transport.

- **Reducing Traffic Congestion and Urban Noise**: Cycling and walking decrease the number of vehicles on the road, which helps reduce congestion and noise levels in cities. This contributes to creating quieter and more livable urban environments.

Encouraging the use of bicycles and walking on short journeys is essential to achieve sustainable urban mobility and promote healthy lifestyles.

Use of Electric and Hybrid Cars

Electric and hybrid cars are more sustainable alternatives to traditional petrol or diesel vehicles, as they produce fewer emissions and can run on renewable energy.

- **Reducing Pollutant Gas Emissions**: Electric cars do not emit greenhouse gases while in operation, and hybrid cars generate fewer emissions than traditional combustion vehicles. This contributes to improved air quality, especially in densely populated urban areas.

- **Charging with Renewable Energy**: Electric cars can be charged with electricity generated from renewable sources, such as solar or wind power. This reduces the vehicle's carbon footprint and decreases dependence on fossil fuels.

- **Less Noise Pollution**: Electric cars are considerably quieter than internal combustion vehicles, which helps reduce noise pollution in cities and improves the well-being of inhabitants.

- **Incentives and Benefits for Sustainable Vehicle Users**: In many countries, governments offer tax incentives and other benefits to encourage the purchase of electric and hybrid cars, such as toll exemption, tax reductions, and the creation of exclusive parking areas.

The shift to electric and hybrid cars is an important option for those who need a private vehicle, and its adoption contributes significantly to reducing emissions from the transport sector.

Promotion of Carpooling and Shared Mobility

Carpooling and shared mobility are alternatives that allow several people to share the same vehicle, which reduces the number of cars on the road and, therefore, carbon emissions. These options are practical and economical, especially for those who need to travel by car.

- **Emission Reduction and Traffic Decongestion**: By carpooling, the number of vehicles on the road is reduced, which decreases emissions per person and helps to decongest traffic in cities. This is especially useful in urban areas with high population density and high traffic.

- **Economic Savings for Users**: Carpooling allows the costs of fuel and vehicle maintenance to be divided among passengers, which makes transportation more accessible and economical. In addition, carpooling platforms make it easier to coordinate rideshares between people who have similar destinations.

- **Fostering a Culture of Sustainable Mobility**: By using carpooling and other shared mobility systems, users join a more sustainable transport culture, which prioritises the efficient use of resources and promotes less dependence on private vehicles.

- **Ride-sharing services and mobility technology**: Mobile apps and digital platforms have made shared mobility more accessible and practical. Ride-sharing companies, such as carsharing or ridesharing, offer on-demand mobility solutions, reducing the need to own a vehicle.

The promotion of carpooling and shared mobility makes it possible to optimise the use of vehicles and contribute to more efficient and sustainable transport.

Teleworking as an Alternative to Reduce Commuting

Teleworking, when feasible, is an effective option to reduce the need to commute daily. Working from home offers multiple environmental and personal benefits, especially in the context of companies and the labor sector.

- **Reducing the Carbon Footprint in Transportation**: By eliminating the need to commute to the workplace, telecommuting reduces the number of vehicles on the roads and, consequently, the CO_2 emissions associated with transportation. This has a positive impact on air quality and reducing fossil fuel consumption.

- **Time Savings and Quality of Life Improvement**: By reducing commuting, teleworking allows employees to save time and improve their quality of life. This reduces the stress associated with commuting and can increase worker satisfaction and well-being.

- **Decreased congestion on public transport and roads**: Teleworking helps to decongest public transport systems and roads, especially at peak times. This allows those who need to travel to do so more quickly and efficiently, also reducing the wear and tear on infrastructure.

- **Adapting to Flexible and Efficient Work Models**: Many companies are adopting hybrid work models that allow teleworking to be combined with face-to-face assistance. This not only improves sustainability, but also allows for greater labor flexibility and efficiency in resource management.

Teleworking represents a sustainable alternative to reduce the environmental impact of transport and improve the quality of life of employees.

Sustainable mobility offers multiple options that each person can adapt to their lifestyle and needs. From using public transport and cycling to shared mobility and teleworking, each alternative contributes to reducing greenhouse gas emissions and improving air quality in our cities. Adopting sustainable mobility practices not only has a positive impact on the environment, but also contributes to people's well-being and health.

Responsible Consumption

Responsible consumption involves adopting habits of purchasing and using products that minimize environmental impact and promote a circular economy. Reduce, reuse and recycle are the three principles that allow each person to make more sustainable choices, optimising the use of resources and reducing waste generation. This approach to consumption helps to reduce our ecological footprint and promotes a more respectful relationship with the environment. Here are some key practices for adopting responsible consumption.

Reduce Consumption of Plastics and Single-Use Products

The use of plastics and single-use products is one of the main environmental problems, as these materials usually take hundreds of years to decompose. Reducing the consumption of these products is essential to protect ecosystems and reduce the amount of waste.

- **Opt for Reusable Products**: Swapping single-use products for reusable alternatives, such as stainless steel bottles, cloth bags, and glass containers, helps reduce the amount of plastics that end up in landfills and oceans. These products, in addition to being more sustainable, are usually more durable and economical in the long term.

- **Purchase of Products with Minimal and Recyclable Packaging**: By choosing products that come in minimal packaging or recyclable materials, such as cardboard or glass, the demand for plastics and single-use materials is reduced. Opting for brands that prioritize sustainable packaging helps foster responsible business practices.

- **Avoid Single-Use Plastics**: Whenever possible, it is advisable to avoid single-use products, such as cutlery, straws, disposable plates and plastic wrapping. These small daily actions help to significantly reduce the amount of waste and promote a culture of conscious consumption.

- **Bulk Purchase and Use of Own Packaging**: Buying food and other products in bulk allows you to reduce the use of plastic containers. By bringing your own containers when making bulk purchases, the use of disposable bags and packaging is minimized, promoting more responsible consumption.

Reducing the consumption of plastics and single-use products not only decreases waste, but also helps protect biodiversity and reduce pollution of natural ecosystems.

Purchase of Local and Seasonal Products

Buying local and seasonal produce is a practice that reduces your carbon footprint and supports the local economy. Seasonal foods and products typically require less energy to produce and transport, making them more sustainable.

- **Reduced Carbon Footprint**: Local and seasonal products require less transportation and lengthy storage, which reduces the carbon footprint associated with their distribution. Food that travels long distances requires additional energy to preserve, which increases CO_2 emissions.

- **Supporting Local Producers**: Buying local products strengthens the community's economy and fosters a more sustainable and resilient supply chain. Supporting local farmers and producers contributes to job creation and reduces dependence on imported food.

- **Fresher and More Nutritious Foods**: Seasonal and local products are usually fresher and retain their nutritional properties better. This is because they do not require preservatives or long-term storage treatments, as is the case with out-of-season foods that must be transported over long distances.

- **Less Reliance on Chemical Inputs**: Seasonal foods are often more resilient to current weather conditions, reducing the need for additional pesticides and fertilizers. This is beneficial for both human health and the environment.

Opting for local and seasonal produce allows consumers to reduce their environmental impact and access fresh, higher-quality produce, while supporting their communities.

Opt for Durable and Quality Products

Investing in good quality products with a long service life helps reduce the need for frequent replacements and reduce waste generation. This practice encourages conscious and more responsible consumption.

- **Long-Term Waste Reduction**: Quality products typically have a long shelf life, which decreases how often they are discarded and replaced. By opting for durable products, waste generation is avoided and the demand for new resources is reduced.

- **Long-Term Economic Savings**: Although high-quality products may have a higher initial cost, they often prove to be more economical in the long run, as they do not require frequent replacements. This applies to a wide range of products, from household appliances to clothing and furniture.

- **Sustainability in Production**: Many companies that produce high-quality items also focus on sustainable production practices, using responsibly sourced materials and minimizing environmental impact. By choosing products from companies committed to sustainability, you support a more responsible economy.

- **Promotion of Conscious Consumption**: Opting for durable products helps to promote a cultural shift towards more conscious consumption and less oriented towards obsolescence. Instead of choosing disposable or short-lived products, this practice promotes the value of durability and sustainability.

Choosing products with good quality and long service life allows the consumer to reduce the amount of waste and enjoy products that require less maintenance and replacement.

Promotion of the Reuse of Materials and Products

Reuse is a fundamental pillar of the circular economy, as it extends the useful life of products and decreases the demand for new resources. Adopting reuse on a day-to-day basis allows you to significantly reduce the amount of waste generated.

- **Buying Second-Hand Clothing and Furniture**: Second-hand clothing and furniture are sustainable options that help reduce the environmental impact of the fashion and furniture industry. By buying second-hand products, you avoid consuming resources for the manufacture of new items.

- **Repair of Products and Appliances**: Repairing appliances and electronic products instead of replacing them contributes to reducing the generation of electronic waste. Repair shops and online guides make it easy to restore damaged products, extending their life.

- **Donation of Unused Objects and Goods**: Instead of discarding objects in good condition, donating them allows other people to take advantage of them. This includes clothing, toys, furniture, and other items that can be reused by people who need them.

- **Reuse of Packaging and Packaging Materials**: Glass containers, cloth bags, and reusable jars can be used to store food and other products. Packaging materials, such as cardboard boxes and bubble wrap, can also be reused to reduce the consumption of new materials.

Reuse extends the useful life of products and reduces pressure on natural resources, fostering a culture of use and sustainability.

Recycling and Waste Separation at Home

Recycling is a key practice in responsible consumption, as it allows materials to be recovered and the amount of waste that ends up in landfills to be reduced. Properly separating waste at home facilitates the recycling process and contributes to more sustainable waste management.

- **Separation of Recyclable Materials**: Separating materials such as paper, glass, plastics, and metals allows them to be recycled effectively. It is important to know your local recycling regulations to ensure that materials are managed correctly and do not contaminate each other.

- **Knowledge of Recycling Symbols and Classification of Plastics**: Familiarizing yourself with recycling symbols and the different types of plastics facilitates the correct disposal of waste. Some plastics, such as PET (polyethylene terephthalate), are easier to recycle, while others, such as PVC, require special treatment.

- **Composting Organic Waste**: Composting is a way to recycle organic waste, such as food scraps and yard waste, into natural compost. This reduces the amount of waste that goes to landfills and generates compost useful for gardening and agriculture.

- **E-Waste Reduction**: E-waste contains materials that can be recycled and reused, such as metals and plastic components. It is advisable to take this waste to specialized recycling points or stores that accept electronic devices for responsible management.

Recycling and waste separation allow materials to be given a second life, reduce the amount of garbage in landfills and promote a more efficient circular economy.

Responsible consumption is a fundamental practice to reduce our environmental impact and live more sustainably. Adopting habits such as reducing plastics, buying local products, choosing durable items, reusing, and recycling not only decreases the amount of waste generated, but also contributes to a circular economy and a more efficient use of natural resources. These habits, applicable in daily life, allow each person to take an active role in protecting the environment and promoting more conscious and sustainable consumption.

Energy Efficiency at Home and Work

Energy efficiency is critical to reducing energy consumption and greenhouse gas emissions, and helps lower costs on your energy bill. Implementing efficiency practices at home and in the work environment allows for a more rational use of resources, while contributing to environmental protection. Below are some of the main strategies to improve energy efficiency in our daily lives.

Use of LED Lighting and Low Energy Consumption Devices

Opting for efficient technologies in lighting and appliances reduces electricity consumption and promotes a more sustainable environment.

- **Energy-Efficient LED Bulbs**: LED bulbs consume up to 80% less energy than traditional incandescent bulbs and have a much longer lifespan, reducing both energy consumption and replacement frequency. LED bulbs are also available in different intensities and colors, allowing the lighting to be adjusted to the needs of the space.

- **Appliances with Energy Efficiency Labels**: Appliances and devices with energy efficiency labels, such as classes A++ or A+++, use less electricity and water. Opting for these appliances reduces the environmental impact and allows you to save on energy consumption in the long term.

- **Automation and Light Sensors**: Installing motion sensors in walkway areas or spaces of occasional use (such as bathrooms, hallways, and garages) allows lights to turn on only when needed and automatically turn off when there are no people in the room. This helps to avoid wasting energy.

- **Dimmable Lamps and Timers**: Dimmable lamps allow you to adjust the intensity of the light according to your needs, while timers and stopwatches make it easy to schedule schedules for the automatic shutdown of lights and devices, further optimizing energy consumption.

Using energy-efficient devices and efficient lighting technologies is one of the most effective ways to reduce electricity consumption at home and at work, without sacrificing comfort.

Natural Light Harvesting and Thermal Insulation

Taking advantage of natural light and properly insulating spaces are simple and effective measures to reduce the use of heating, air conditioning, and artificial lighting systems.

- **Use of Natural Light in the Home and Office**: Placing workstations near windows and making the most of sunlight during the day decreases the need for artificial lighting. Decorating with light colors also helps to reflect natural light, increasing the luminosity in interior spaces.

- **Natural Ventilation and Air Flow Control**: Cross ventilation and opening windows at strategic times allow you to regulate the temperature in the home without the need for air conditioning. The use of curtains or blinds also helps to control the entry of heat in summer and to maintain the temperature in winter.

- **Insulation of Windows and Doors**: Good thermal insulation in windows, doors and walls maintains the temperature in the home, reducing the need for heating in winter and air conditioning in summer. Sealing air leaks and using double glazing on windows are measures that can improve the thermal efficiency of the space.

- **Use of Thermal Blinds and Curtains**: Thermal blinds and blinds help to keep heat in winter and block heat from entering in summer, reducing the use of air conditioning systems. These measures can also improve privacy and reduce noise in the home or office.

Taking advantage of natural light and properly insulating spaces is a practical way to reduce energy consumption and maintain a comfortable temperature throughout the year.

Shutdown of Devices on Standby

Electronic devices that remain in standby mode constantly consume power, even when not in use. This "phantom" energy accounts for a significant percentage of total consumption in many homes and offices.

- **Disconnecting Electronic Devices**: Disconnecting devices such as televisions, computers, microwaves and chargers when they are not in use avoids standby power consumption. It is advisable to unplug them or use automatic shut-off strips that cut off the electricity supply when they are not needed.

- **Smart and Automatic Power Strips**: Smart power strips allow you to schedule the shutdown of connected devices or turn them off remotely. There are also automatic shut-off strips that detect when a device is not in use and cut off the power supply autonomously.

- **Disabling Unnecessary Features on Electronic Devices**: Many devices include features that are automatically activated in standby mode, such as software updates or Wi-Fi settings. Adjusting device settings so that they don't consume standby power helps reduce power consumption.

- **Using Timers for Appliances**: Timers and stopwatches are useful for limiting the running time of certain devices, such as water heaters and ventilation systems, ensuring that they are only used when they are actually needed.

Turning off or unplugging devices in standby mode is a simple but effective measure to reduce energy consumption at home and at work, and helps to avoid wasting electricity.

Rational Use of Heating and Air Conditioning

The use of heating and air conditioning systems represents one of the largest sources of energy consumption in the home and in the work environment. Regulating their use efficiently helps reduce environmental impact and energy costs.

- **Comfort temperatures in winter and summer**: In winter, it is recommended to adjust the heating to about 20 °C and in summer the air conditioning to about 24 °C. These temperatures are sufficient to maintain comfort and avoid excessive energy consumption. Each additional grade can increase energy consumption by 7%.

- **Using Smart Thermostats**: Programmable thermostats and smart thermostats allow you to regulate the temperature automatically based on schedules or usage patterns. This ensures that heating and air conditioning only work when necessary, optimizing energy consumption.

- **Fans and Air Circulation Systems**: Fans are an efficient and energy-efficient option for regulating temperature in summer. Using ceiling fans in combination with the air conditioner allows fresh air to be distributed and the load on the HVAC system to be reduced.

- **Thermal Insulation and Air Flow Control**: As mentioned above, thermal insulation is essential to preserve the interior temperature, preventing heat loss in winter and heat entry in summer. This reduces the need to overuse heating or air conditioning.

Rational use of heating and air conditioning not only helps to save energy, but also improves comfort at home and in the workplace.

Renewable Energy Implementation

The installation of renewable energy systems, such as solar panels or geothermal heating systems, makes it possible to take advantage of clean energy sources and reduce dependence on fossil fuels. Although the initial investment can be high, these systems offer long-term savings and contribute to sustainability.

- **Solar Panels for Electricity Generation and Heating**: Photovoltaic and thermal solar panels are an increasingly accessible option for generating electricity and hot water at home and at work. These systems take advantage of solar energy, a clean and renewable source, to cover a significant part of the energy needs.

- **Geothermal and Aerothermal Heating**: Geothermal and aerothermal heating are air conditioning systems that take advantage of energy from the subsoil or air to heat or cool indoor spaces. These technologies are highly efficient and can significantly reduce energy consumption.

- **Installing Wind Turbines in Rural Environments**: In rural areas or open spaces, small-scale wind turbines allow electricity to be generated by wind. These facilities are suitable for locations where wind conditions are favorable and can complement other renewable energy sources.

- **Economic and Ecological Benefits of Renewable Energies**: In addition to savings on electricity bills, renewable energies reduce CO_2 emissions and dependence on fossil fuels, contributing to the protection of the environment. In many countries, governments offer incentives and subsidies to facilitate the adoption of renewable energy in households and businesses.

The implementation of renewable energies is a long-term investment that allows people to reduce their environmental impact and contribute to a more sustainable energy system.

Energy efficiency is an accessible and practical strategy that every person can adopt at home and at work to reduce energy consumption and greenhouse gas emissions. From using efficient devices and improving insulation to adopting renewable energy, every action contributes to sustainability and saves on energy consumption. These practices not only protect the environment, but also offer economic benefits, promoting a more balanced and responsible life.

Sustainable Food

Food is one of the aspects that has the greatest impact on the environment, from the use of natural resources to the greenhouse gas emissions generated in its production. Adopting a sustainable diet not only helps to reduce our ecological footprint, but also contributes to the conservation of ecosystems and biodiversity. Here are some eating habits that can make a big difference in terms of sustainability.

Reducing Consumption of Meat and Animal Products

The production of meat and animal products is one of the main sources of greenhouse gas emissions and has a significant impact on natural resources. Adopting a more plant-based diet helps reduce these emissions and protect ecosystems.

- **Environmental Impact of Meat Production**: Livestock farming is responsible for a large portion of methane emissions, a potent greenhouse gas, due to the digestion of ruminant animals. In addition, meat production often requires large tracts of land, which contributes to deforestation, especially in regions such as the Amazon, where forests are destroyed to create pastures.

- **Benefits of a Diet with More Plant Foods**: Opting for foods such as legumes, grains, and vegetables instead of animal products not only reduces carbon emissions, but also decreases the demand for natural resources, such as water and land. Plant-based diets typically have a smaller ecological footprint and can be just as nutritious.

- **Plant-based alternatives**: There are many plant-based protein options today, such as tofu, tempeh, and legumes, which are sustainable alternatives to meat. You can also find plant-based products that simulate the texture and taste of meat, for those looking for more eco-friendly options without giving up their favorite foods.

- **Encouraging Responsibly Produced Dairy and Meat Consumption**: When consuming meat or dairy products, opting for responsible production options, such as grass-fed meat or organic dairy, helps reduce environmental impact and supports more sustainable livestock practices.

Reducing the consumption of meat and animal products is one of the most effective actions to reduce the carbon footprint of our diet and promote a more rational use of resources.

Buying Organic and Sustainable Agriculture Products

Organic and sustainably farmed products are grown without pesticides or chemical fertilizers, which contributes to the health of the soil, water and ecosystems. Buying these products supports farmers who promote responsible practices and helps protect biodiversity.

- **Benefits of Organic Food**: Organic food is produced without the use of synthetic chemicals, which reduces soil and water pollution. By not using pesticides, organic products promote biodiversity and protect pollinators, such as bees, which are essential for food production.

- **Supporting Sustainable and Regenerative Agriculture**: Sustainable and regenerative agriculture focuses on maintaining and improving soil fertility, conserving water, and protecting local ecosystems. By supporting farmers who use these practices, you contribute to the creation of a more resilient food system in harmony with the environment.

- **Organic and Sustainable Farming Certifications**: In many countries, products that meet organic standards have certifications that guarantee that they have been grown in an environmentally friendly way. Opting for products with these certifications is one way to support sustainable production.

- **Increased Demand and Accessibility for Eco-Friendly Products**: As more people opt for eco-friendly products, the demand increases, making it easier for these products to become more accessible and available. This also incentivizes other farmers to adopt more sustainable practices.

Buying organic and sustainably farmed products not only benefits the environment, but also supports local farmers and promotes a healthier and more balanced food system.

Reducing Food Waste

Food waste is one of the main environmental problems, as discarded food generates greenhouse gases by decomposing in landfills. Reducing food waste is one of the most effective ways to decrease our ecological footprint.

- **Meal Planning**: Meal planning and making a shopping list helps avoid overbuying and reduce food waste. By having a clear plan, it is less likely that unnecessary food will be purchased that may end up expiring.

- **Food Storage and Preservation**: Properly storing food, using methods such as refrigeration, freezing and the use of airtight containers, allows you to prolong its freshness and prevent it from spoiling. This also helps to make the most of food and reduce waste.

- **Utilization of Leftovers**: Leftovers can be used to create new recipes, thus avoiding waste. For example, leftover vegetables can be made into soups or stews, while stale bread can be used to make toast or puddings.

- **Composting Organic Waste**: When food cannot be used, composting is a sustainable alternative to managing organic waste. Composting transforms food scraps into natural compost, which can be used to enrich the soil and reduce the amount of waste that ends up in landfills.

Reducing food waste contributes to reducing the amount of waste and making better use of the natural resources invested in food production.

Use of Products with Lower Environmental Impact

Some food products require large amounts of water and resources to produce. Opting for foods with a lower environmental impact helps to conserve natural resources and reduce the ecological footprint of our diet.

- **Opt for Seasonal and Local Foods**: Seasonal and local foods typically have a lower environmental impact, as they do not require long-distance transportation or prolonged storage. These foods are often fresher and more nutritious, and contribute to a more sustainable local economy.

- **Foods with a Low Water Footprint**: Some foods, such as rice and avocados, have a high water footprint, that is, they require large amounts of water for their production. Opting for foods with a lower water footprint, such as seasonal vegetables and fruits, helps conserve water resources, especially in drought-affected regions.

- **Minimize the Consumption of Processed Products**: Ultra-processed foods usually require a large amount of energy and resources to prepare and package. By opting for fresh and minimally processed foods, the environmental impact associated with their production is reduced.

- **Avoid Consumption of Exotic or Imported Foods**: Foods that require international transportation, such as some exotic products, generate a higher carbon footprint due to the long distances they must travel. Whenever possible, it is preferable to consume local products to minimize environmental impact.

Choosing foods with a lower environmental impact contributes to the sustainability of natural resources and promotes more conscious and respectful consumption.

Conscious and Efficient Cooking

The way we cook also influences our ecological footprint. Adopting efficient cooking practices and using sustainable utensils helps reduce energy consumption and minimize environmental impact.

- **Use of Pressure Cookers and Induction Cookers**: Pressure cookers allow cooking in less time, which reduces energy consumption. Induction cooktops are also an efficient alternative, as they heat faster and more evenly compared to conventional gas or electric cooktops.

- **Cover pots and utensils**: Covering pots while cooking helps to conserve heat and reduce cooking time. This allows for considerable energy savings, especially when preparing dishes that require long cooking times.

- **Use of Sustainable Cookware**: Opting for durable cookware, such as stainless steel, glass, and wood, reduces the need for frequent replacement and avoids waste. These materials are more resistant and tend to have a lower environmental impact.

- **Preparing Recipes That Make the Most of Food**: Mindful cooking involves using all parts of food, such as vegetable peels to make broths or seeds to add to salads. This minimizes waste and allows you to make the most of nutrients.

Cooking efficiently not only reduces energy consumption, but also allows you to enjoy healthier food and optimize resources at home.

Adopting a sustainable diet is an effective way to reduce our environmental impact and contribute to the conservation of natural resources. From reducing meat consumption and supporting organic farming to minimizing food waste and choosing products with lower impact, every action counts. In addition, sustainable food promotes a more balanced and healthy diet, benefiting both the environment and our health.

Responsible Use of Water

Water is one of the most essential and finite natural resources, and its responsible consumption is essential to ensure its availability in the future. The water crisis affects many regions of the world, and with climate change, water scarcity is an increasing concern. Adopting efficient water saving and management practices at home and in the community is an effective way to contribute to sustainability. Below, various strategies are explored to use water responsibly and reduce its waste.

Reducing Water Consumption in the Home

The home is one of the main places where water consumption can be optimized. Adopting small changes in our daily habits can have a big impact on saving water.

- **Shorter, More Efficient Showers**: Reducing shower time helps conserve a significant amount of water. Installing low-flow showerheads also allows water consumption to be decreased without compromising the quality of the shower.

- **Turn Off the Faucet When Brushing Your Teeth or Everyday Tasks**: Turning off the faucet while brushing your teeth, washing your hands, or scrubbing dishes helps save water. These moments represent liters of water wasted if the tap remains unnecessarily open.

- **Repairing Leaks in Pipes and Faucets**: Water leaks may seem insignificant, but they represent a large loss of water over time. Fixing leaky faucets or pipes is a simple and effective measure to avoid waste.

- **Use of Energy-Efficient Toilets**: Low-water toilets use less water in each flush. There are also dual discharge systems, which allow a minimum amount of water to be used for liquid waste and a greater amount for solid waste, optimizing water consumption.

Reducing water consumption in the home is one of the most accessible and effective ways to preserve this valuable and necessary resource.

Using Efficient Appliances

Household appliances such as washing machines and dishwashers consume large amounts of water and energy. Opting for efficient models and using them rationally helps to significantly reduce water consumption.

- **Energy-efficient washing machines and dishwashers**: Energy-efficient appliances are designed to use fewer resources in each wash cycle. These models not only help conserve water, but also reduce energy consumption, benefiting the environment and saving on electricity and water bills.

- **Efficient Use of Appliances**: It is advisable to use washing machines and dishwashers only when they are completely full, to maximize the efficiency of each load. Avoiding extra cycles and using short wash programs also helps reduce water consumption.

- **Cold or Low Temperature Wash**: Many appliances have the option of cold or low temperature washing, which allows you to reduce energy consumption. This fit, in addition to being more sustainable, is beneficial for extending the life of the garments.

- **Regular Appliance Maintenance**: Performing proper maintenance, such as cleaning filters and checking connections, ensures that appliances are running efficiently and prevents water waste caused by clogs or mechanical problems.

Efficient appliances and their rational use allow you to considerably reduce water consumption and contribute to a more sustainable home.

Water Reuse at Home

Water reuse is a practical and sustainable measure that makes it possible to make the most of this resource. There are several ways to reuse water at home, especially in cleaning and watering tasks.

- **Rainwater Collection for Irrigation**: Collecting rainwater in containers is a natural and free way to have water available for the irrigation of gardens, plants and orchards. This practice is especially useful in times of drought and reduces dependence on drinking water for gardening tasks.

- **Using Water from Washing Machines for Exterior Cleaning**: The water used in the washing machine wash cycle can be reused to clean exteriors, such as patios, entrances or terraces. It is important to make sure that the water does not contain harsh or toxic chemicals.

- **Reuse of Cooking Water for Irrigation**: The leftover water from cooking vegetables or pasta can be cooled and used to water plants, as long as it does not contain salt or other additives. This type of water contains nutrients that are beneficial for plant growth.

- **Shower Water Collection**: During the time when the water is heated at the start of the shower, containers can be placed to collect that water and reuse it for other tasks, such as cleaning the home or watering plants.

Reusing water at home is a simple and effective measure to take advantage of this resource and reduce the demand for drinking water.

Sustainable Irrigation and Low Water Consumption Plants

Irrigation is one of the activities that consumes large amounts of water, especially in gardens and orchards. Adopting efficient irrigation techniques and choosing suitable plants helps conserve water and maintain green spaces responsibly.

- **Drip Irrigation and Automated Irrigation Systems**: Drip irrigation is a method that allows water to be supplied directly to the root of plants, avoiding waste and optimizing consumption. Automated irrigation systems can be programmed to operate at specific times, avoiding overwatering or at times of high evaporation.

- **Choosing Native and Drought-Resistant Plants**: Native plants and drought-resistant species are adapted to the local climate and require less water to grow. These plants are ideal for sustainable gardens, as they need less watering and are more resistant to extreme weather conditions.

- **Using Mulch to Conserve Moisture**: Applying a layer of mulch or mulch to the soil around the plants helps retain moisture, reduce evaporation, and protect the roots. This practice reduces the need for irrigation and improves soil health.

- **Watering in Hours of Low Evaporation**: Watering early in the morning or at dusk minimizes evaporation and allows plants to absorb more water. This habit is especially important in hot weather, when water loss through evaporation is greater.

Sustainable irrigation and the right choice of plants can conserve water and maintain healthy gardens, even in conditions of water scarcity.

Virtual Water Consumption Awareness

Virtual water is the water used in the production of goods and food that we consume, from food products to clothing. Being aware of virtual water consumption and reducing the demand for products with a high water footprint is an indirect way to conserve water.

- **Reducing the Consumption of High Water Footprint Products**: Some products, such as meat, coffee, and cotton, require large amounts of water for their production. Opting for alternatives with a lower water footprint, such as plant-based foods, or choosing textiles made of sustainable materials helps to reduce the impact on water resources.

- **Responsible Purchasing of Clothing and Textile Products**: The textile industry is one of the most water-intensive industries. Reducing the purchase of new clothes, opting for second-hand garments and choosing sustainable materials (such as linen or hemp) contributes to more responsible water consumption.

- **Water Savings in Food Production**: By opting for a more plant-based diet, the virtual water consumption associated with food production is reduced. Animal products often require more water than plant foods, so a balanced plant-based diet also contributes to water sustainability.

- **Conscious and Local Consumption**: Local products tend to have a smaller virtual water footprint, as they do not require long-distance transport or preservation processes. Consuming local products reduces the demand for water and supports the local economy.

Being aware of virtual water consumption reduces the indirect impact on water resources and fosters a more respectful relationship with water.

The responsible use of water is essential to ensure the sustainability of this essential resource. From reducing household consumption and using efficient appliances to water reuse and virtual water awareness, every measure contributes to the conservation of water resources. Adopting habits of saving and efficient management of water is not only beneficial for the environment, but it is also a responsibility we share to ensure its availability for future generations.

THE COMPANY TOWARDS SUSTAINABILITY

Sustainability is a growing priority in the business world. Rather than just making financial gains, more and more companies are integrating sustainable practices into their operations and redefining the concept of success. This transformation involves adopting a triple impact approach that considers environmental, social and economic well-being. Below are the main areas where companies can work to lead the shift towards a sustainable business model.

Redefining Business Success in Terms of Sustainability

The concept of business success has evolved to incorporate sustainability as one of its fundamental pillars. Beyond the financial benefits, companies are recognizing that their environmental and social impact is crucial to their long-term sustainability. Integrating sustainability into business objectives allows companies to create positive change in the world, generating value for all their stakeholders. Below are the key elements for redefining business success in terms of sustainability.

Triple Impact: Environmental, Social and Economic

Triple impact refers to the integration of environmental, social, and economic goals into business operations, recognizing that success must benefit not only shareholders, but also communities and the planet.

- **Environmental Impact: Protection of Natural Resources**: Sustainable companies implement practices that reduce the consumption of natural resources, minimize carbon emissions, and promote biodiversity. For example, reducing the carbon footprint and energy efficiency are key objectives for companies committed to positive environmental impact.

- **Social Impact: Commitment to Community Well-Being**: Social impact involves companies improving the quality of life for their employees and communities. This includes practices such as fair working conditions, diversity and inclusion, and community development programs that benefit the areas in which they operate.

- **Economic Impact: Profitability with Responsibility**: Profitability is still critical, but with sustainability, companies seek to achieve profits that are aligned with their environmental and social responsibility. This involves adopting sustainable business models that generate long-term benefits without compromising future resources.

Triple impact allows companies to generate value on multiple levels and ensure sustainable growth that benefits all actors involved.

Valuation of Natural Resources as Assets

Companies that value natural resources recognize that they are essential and limited assets. The protection and rational use of natural resources not only benefit the environment, but also strengthen the company's long-term sustainability.

- **Waste Reduction and Circular Economy**: Sustainable companies implement circular economy models to reduce the waste of materials and resources. This involves adopting recycling, material reuse, and product redesign practices to minimize waste and extend the life of resources.

- **Optimizing Water and Energy Use**: Water and energy are finite resources that must be managed efficiently. Companies can implement measures such as water recycling, the use of renewable energy, and energy efficiency in their facilities to reduce their environmental impact and reduce operating costs.

- **Biodiversity Protection and Ecosystem Conservation**: The preservation of biodiversity and ecosystems is fundamental to sustainability. Companies can collaborate with reforestation projects, conservation of natural habitats and recovery of endangered species as part of their commitment to the environment.

- **Investment in Technology for Resource Efficiency**: Technology plays a crucial role in optimizing the use of resources. Companies can invest in precision technologies, such as water and energy consumption monitoring, to improve efficiency and reduce environmental impact.

Valuing natural resources as assets allows companies to protect the natural environment in which they operate and ensure their availability for the future.

Innovation as a Driver of Sustainability

Sustainability is not only a responsibility, but also an opportunity to innovate. Companies that integrate sustainability into their strategies drive creativity and the development of solutions that minimize their environmental impact and improve their competitiveness.

- **Sustainable and Recyclable Product Design**: Innovation in product design allows companies to create sustainable and recyclable products, reducing their environmental impact from the production phase. This includes using recycled, biodegradable materials and making products that are easily recyclable at the end of their useful life.

- **Adoption of Clean Energy Technologies**: Investing in renewable energy, such as solar and wind power, allows companies to reduce their dependence on fossil fuels and decrease their greenhouse gas emissions. In addition, the use of energy storage technologies helps to manage energy demand efficiently.

- **Implementation of Circular Economy Solutions**: The circular economy encourages the creation of business models that reuse products and materials, extending their useful life and reducing waste. This includes recycling production waste and reusing materials in other processes, optimizing resources and reducing waste.

- **Use of Technology for Process Efficiency**: Innovation in process management, such as the use of artificial intelligence for production optimization, allows for reduced energy consumption and improved operational efficiency. This technology allows companies to analyze and optimize their processes in real time, reducing environmental impact.

Sustainability drives innovation and enables companies to adapt to an ever-changing market environment, creating solutions that benefit both the business and the planet.

Business Ethics and Transparency

Business ethics and transparency are critical for companies to gain the trust of their stakeholders and stay accountable for their sustainable practices. Accountability and clear communication are essential in the age of sustainability.

- **Transparent Communication of Sustainable Goals**: Sustainable companies communicate their sustainability goals and progress made in a transparent manner. This allows consumers, employees, and shareholders to be informed about the company's commitments and monitor its progress.

- **Accountability through Sustainability Reporting**: Publishing sustainability reports allows companies to assess their environmental and social performance and communicate their achievements and challenges. Sustainability reports should include specific data, such as energy consumption, CO_2 emissions, and social impact projects, to demonstrate commitment to sustainability.

- **Fostering an Ethical and Responsible Culture**: Business ethics begins with an organizational culture that values sustainability and responsibility. Companies must establish clear codes of conduct and offer training to their employees to foster a culture of respect and care for the environment and society.

- **Alignment with the Principles of the Sustainable Development Goals (SDGs):** The UN SDGs provide guidance for businesses to adopt ethical and responsible practices. Companies can align their goals with the SDGs to demonstrate their commitment to global issues such as gender equality, poverty reduction, and climate action.

Ethics and transparency in business practices strengthen credibility and trust, allowing companies to build lasting relationships with their stakeholders and foster a fairer and more sustainable market.

Redefining business success in terms of sustainability allows companies to achieve a broader vision that includes environmental, social and economic well-being. By taking a triple impact approach, valuing natural resources, innovating products and processes, and prioritizing ethics and transparency, companies can lead the way to a more sustainable future. This approach not only generates benefits for the company, but also creates a positive impact on society and the planet, allowing business success to be synonymous with responsibility and sustainability.

Sustainable Companies and their Best Practices

Globally, some companies are taking the lead in the transformation towards a more sustainable business model. These examples show that sustainability is not only feasible, but also beneficial for the company, the environment, and society. Below, the sustainability practices implemented by prominent companies in different sectors are explored.

Patagonia: Innovation and Environmental Responsibility

Patagonia is an outdoor clothing and equipment company that has taken an approach focused on sustainability and environmental responsibility. Over the years, the company has implemented numerous initiatives that have made it a benchmark in sustainability in the fashion sector.

- **Use of Recycled and Organic Materials**: Patagonia uses recycled and organic materials in the manufacture of its products. By reducing reliance on virgin materials, the company minimizes its environmental impact, decreasing the use of natural resources and greenhouse gas emissions.

- **Promoting Product Repair and Reuse**: The company encourages its customers to repair and

210

reuse their products instead of buying new ones. Its "Worn Wear" program facilitates the repair and sale of used clothing, promoting a circular economy in which products have a longer useful life.

- **Commitment to Environmental Causes**: Patagonia donates a percentage of its profits to environmental organizations and actively promotes conservation causes. The company has carried out campaigns to raise awareness about climate change and the protection of ecosystems, involving both its employees and its customers.

- **Transparency in the Supply Chain**: The company maintains absolute transparency in its production practices, reporting on the working conditions in its factories and the sustainable practices of its suppliers.

Patagonia proves that sustainability can be integrated into every aspect of a company, creating value for both customers and the environment.

IKEA: Transition to a Circular Economy

IKEA, one of the largest furniture companies globally, is committed to sustainability and has adopted a circular economy approach to reduce its environmental impact and promote recycling and reuse of products.

- **Reusable and Recyclable Product Design**: The company has redesigned many of its products so that they can be reused, recycled or composted. This includes the use of sustainable materials and designs that make it easy to disassemble and repair.

- **Recycling and Waste Reduction Programs**: IKEA has implemented recycling programs in its stores so that customers can return furniture and recycle materials. These programs are designed to reduce waste and facilitate a circular economy.

- **Furniture Rental**: In some markets, IKEA has begun offering furniture rental services, allowing customers to access high-quality products without the need to purchase them. This makes it easier to reuse and extends the shelf life of the products.

- **Commitment to Renewable Energy**: IKEA has invested in solar and wind energy to reduce its carbon footprint. It aims to exclusively use renewable energy throughout its operations and has installed solar panels in many of its stores and distribution centers.

IKEA proves that a multinational company can transition to a sustainable business model and promote responsible practices throughout its supply chain.

Unilever: Commitment to Sustainability in its Supply Chain

Unilever, one of the largest manufacturers of consumer goods, has adopted a comprehensive sustainability strategy that encompasses its supply chain, its products and its social impact.

- **Reducing Plastic Use**: Unilever has set ambitious targets to reduce the use of plastic in its products and packaging. The company has committed to making 100% of its plastic packaging reusable, recyclable or compostable by 2025.

- **Promoting Sustainable Agriculture**: The company works with farmers in several countries to implement sustainable agriculture practices, such as crop rotation and reducing pesticide use. This helps conserve soil and protect biodiversity.

- **Improving Supplier Conditions in Developing Countries**: Unilever works with its suppliers to improve labour and social conditions in their supply chains. This includes paying fair wages and improving workplace safety, especially in developing countries.

- **Transparency and Communication of Sustainability Progress**: The company regularly publishes its sustainability progress in publicly accessible reports, which fosters transparency and brand trust.

Unilever has shown that a large consumer goods company can lead the shift towards sustainability and promote responsible practices throughout its supply chain.

Tesla: Innovation in Sustainable Energy and Transportation

Tesla, a leader in the manufacture of electric vehicles and clean energy technology, has revolutionized the automotive industry and promoted the adoption of renewable energy.

- **Electric Vehicles to Reduce CO_2 Emissions**: Tesla has focused on producing high-performance electric vehicles that help reduce greenhouse gas emissions compared to gasoline and diesel vehicles. This innovation has made electric vehicles increasingly accessible and attractive to consumers.

- **Energy Solutions for Homes and Businesses**: In addition to electric vehicles, Tesla has developed solar panels and batteries for energy storage. These solutions allow homes and businesses to generate and store clean energy, reducing their dependence on the electricity grid.

214

- **Advances in Battery Technology**: The company has invested in the development of batteries with longer life and capacity, which improves the range of its vehicles and the efficiency of its energy solutions. This innovation is key to making electric vehicles and renewables more viable on a large scale.

- **Commitment to Expanding Charging Infrastructure**: Tesla has installed a network of fast-charging stations to facilitate the use of electric vehicles around the world. This infrastructure is essential to promote the mass adoption of electric vehicles and reduce barriers to access for consumers.

Tesla has been a pioneer in promoting sustainable transportation and energy, driving structural change in one of the most polluting industries.

Danone: Sustainability in the Food Industry

Danone, an international dairy and food company, has implemented sustainability practices in all aspects of its business and is committed to a B Corp model, which prioritizes social and environmental impact.

- **Regenerative Agriculture and Responsible Practices**: Danone works with farmers to implement regenerative agriculture practices, which promote soil health, reduce carbon emissions, and foster biodiversity. This includes techniques such as using cover crops and reducing chemicals in crops.

- **Reduction of Sustainable Plastics and Packaging**: The company has adopted measures to reduce the use of plastics and encourage recycling in its packaging. Danone has begun to introduce biodegradable and recyclable packaging in its products, aligning with its commitment to reduce its environmental impact.

- **Animal Welfare and Social Sustainability**: Danone promotes animal welfare in its livestock practices and collaborates with local producers to ensure ethical conditions. In addition, the company has support programs for its suppliers in vulnerable communities, improving their working and economic conditions.

- **Certification as a B Corp**: Danone has adopted the B Corp model, a certification that assesses the social and environmental impact of the company. This implies a commitment to transparency and regular evaluation of its sustainable performance.

Danone demonstrates that sustainability and profitability can coexist in the food sector, promoting practices that benefit both the environment and communities.

These examples of sustainable businesses show that implementing sustainability practices is not only feasible, but also cost-effective and beneficial for everyone involved. Patagonia, IKEA, Unilever, Tesla and Danone are leading the change in their sectors, demonstrating that sustainability is an added value that attracts consumers, improves brand reputation and contributes to a more balanced world. These companies are an inspiration to other organizations and demonstrate that sustainability is a path to long-term business success.

Environmental Education in the Workplace

Environmental education in the workplace is a key component to fostering a culture of sustainability in the company. By training employees in responsible practices, you not only reduce the organization's environmental impact, but also promote a sense of collective responsibility. This shared commitment creates a more cohesive and responsible work environment, where employees are aware of the role they play in caring for the planet. Below are effective strategies for implementing comprehensive environmental education in the workplace.

Sustainability Training Programs

Sustainability training programs train employees in practical knowledge that allows them to understand and apply sustainable practices in their daily activities.

- **Energy Efficiency and Waste Reduction Trainings**: These trainings teach employees how to minimize energy consumption in their daily activities and how to manage waste responsibly. Sessions can include energy-saving strategies, such as turning off equipment and lights, and methods for reducing, reusing, and recycling materials.

- **Workshops on the Sustainable Use of Resources**: Conducting workshops that explain how to manage resources such as paper, water and office supplies helps to reduce the environmental impact in the day-to-day of the company. These workshops can range from efficient paper use to water-saving practices in the workplace.

- **Carbon Footprint Awareness**: A program that includes carbon footprint information helps employees understand how their actions contribute to greenhouse gas emissions and what they can do to reduce them. This type of training can include calculating your personal carbon footprint and strategies to decrease it both at work and at home.

- **Corporate Social Responsibility (CSR) training**: CSR training programs allow employees to learn about the company's commitment to sustainability and how their participation contributes to achieving the organization's CSR goals. Training in this topic strengthens employees' sense of belonging and commitment to the company's values.

Implementing sustainability training programs allows employees to become agents of change, applying sustainable practices inside and outside of work.

Recycling and Waste Reduction Initiatives

Encouraging recycling and waste reduction in the workplace is a tangible and effective way to reduce the organization's environmental impact.

- **Establishment of Recycling Points**: Installing clearly identified recycling stations for different types of waste, such as paper, plastic and organics, facilitates recycling in the workplace. In addition, including informative signs on how to separate waste correctly helps improve the effectiveness of recycling.

- **Reducing Single-Use Materials**: Limiting the use of disposable materials, such as plastic cups, cutlery, and plates, is an important measure. Companies can choose to offer reusable tableware in canteens and kitchens, and provide reusable water bottles for employees.

- **Recycling of Electronic Equipment and Waste Electrical and Electronic Equipment (WEEE):** Establishing a collection point for electronic equipment, such as computers or printers that are no longer in use, allows this waste to be properly managed. Obsolete equipment can be donated or recycled in specialized plants.

- **Paper Reduction Programs**: Encouraging the use of digital tools and minimizing paper printing reduces the amount of waste generated. Companies can establish responsible printing policies and encourage employees to work digitally whenever possible.

- **Waste Reduction Challenges**: Conducting regular waste reduction challenges, such as a contest to reduce the use of paper or plastics, motivates employees to adopt sustainable habits and makes the process more engaging and participatory.

These initiatives create a culture of waste reduction at work and promote a cleaner and more sustainable environment.

Promoting Sustainable Mobility Among Employees

Promoting sustainable mobility is one way to reduce the company's carbon footprint and facilitate more responsible transportation options for employees.

- **Bicycle Incentives**: Companies may offer incentives for those who choose to use bicycles as a means of transportation, such as reimbursement for bicycle repair expenses or the installation of bicycle parking lots. Some workplaces also offer showers and changing rooms for employees commuting by bicycle.

- **Carpooling and Ride-Sharing**: Making it easier to organize carpooling among employees who live nearby allows you to reduce the number of vehicles on the road and therefore CO_2 emissions. Companies can implement internal platforms to coordinate ride-sharing.

- **Support for the Use of Public Transport**: Some companies offer subsidised public transport cards or discounts on transport passes, encouraging the use of public transport. This measure reduces traffic and pollution in cities and contributes to improving air quality.

- **Teleworking and Flexible Schedules**: Teleworking and flexible schedules reduce the need for daily commutes, which contributes to reducing the company's carbon footprint. This measure also improves the quality of life of employees and reduces congestion at peak times.

- **Electric Vehicle Charging Stations**: Installing electric vehicle charging stations in the workplace is an incentive for employees to opt for this transportation alternative. Facilitating access to charging infrastructure contributes to the adoption of sustainable vehicles.

Promoting sustainable mobility is a measure with benefits for both the environment and employees, who can enjoy healthier and more responsible transport options.

Promoting Energy Efficiency in the Workplace

Energy efficiency in the workplace not only reduces operating costs, but also minimizes the company's environmental impact. Educating employees on the rational use of energy is an essential practice to promote sustainability.

- **Awareness of Turning Off Equipment and Lights**: Encouraging employees to turn off lights and equipment when not in use is a simple but effective practice to reduce energy consumption. Reminder posters and training sessions can reinforce this habit.

- **Use of Energy Efficiency Technologies**: Installing motion sensors for lighting, energy-efficient HVAC systems, and energy-efficient equipment in offices contributes to the sustainability of the company. These systems optimize energy use and reduce waste.

- **Encouraging the Use of Renewable Energy**: Whenever possible, companies can opt for renewable energy sources, such as solar or wind power, to reduce their dependence on fossil fuels. In addition, some organizations install solar panels on their premises to cover part of their energy consumption.

- **Reducing Dependence on HVAC Systems**: Adopting passive HVAC measures, such as the use of natural ventilation and optimizing sunlight, reduces reliance on air conditioning and heating systems. These practices, in addition to being sustainable, improve comfort in the workplace.

- **Energy Saving Contests**: Organizing energy saving contests between different departments or teams promotes friendly competition and encourages energy savings in the company. These contests can also serve to reward and recognize the efforts of employees.

Energy efficiency at work is key to reducing the carbon footprint and creating a work environment committed to saving and respecting the environment.

Sustainability Groups and Environmental Ambassadors

Engaging employees in sustainability initiatives through sustainability groups and environmental ambassadors creates a culture of involvement in environmental issues.

- **Creation of Sustainability Groups**: Forming sustainability groups in the company allows employees to participate in the planning and execution of sustainable initiatives. These groups may meet regularly to come up with new ideas and evaluate the progress of environmental initiatives in the company.

- **Appointment of Environmental Ambassadors**: Appointing environmental ambassadors in each department or team motivates employees to lead sustainable initiatives and act as role models for their colleagues. Ambassadors can host awareness events, answer questions about sustainability, and promote environmentally friendly practices.

- **Awareness and Volunteering Events**: Organizing awareness events, such as workshops, talks, and volunteer activities in collaboration with environmental organizations, fosters a sense of responsibility in the team. Activities such as cleaning up natural spaces or planting trees allow employees to actively participate in protecting the environment.

- **Employee Surveys and Feedback**: Conducting surveys to find out employees' opinions and ideas on sustainability allows you to improve sustainability initiatives and adapt strategies according to their suggestions. This participatory approach creates an inclusive work environment focused on continuous improvement.

- **Sustainability Recognitions and Awards**: Recognizing and rewarding employees and teams that demonstrate a commitment to sustainability is an effective way to reinforce the culture of sustainability in the company. These recognitions may include symbolic awards, financial incentives, or prominent mentions in the company's internal newsletters.

The creation of sustainability groups and the appointment of environmental ambassadors foster a work environment in which sustainability is a shared priority and promote change towards more responsible business practices.

Environmental education in the workplace is essential to building an organizational culture committed to sustainability. From training and waste reduction programs to mobility and energy efficiency initiatives, every practice contributes to a more environmentally friendly work environment. In addition, sustainability groups and environmental ambassadors encourage employee participation, making everyone part of the change to a more sustainable future. Environmental education in the company is a powerful tool to reduce environmental impact, promote collective responsibility and improve the reputation of the organization in an increasingly conscious market.

Towards Circular Economy Models

The circular economy is an approach in which products and resources are kept in use for as long as possible, extracting maximum value before they reach the end of their useful life. This model contributes to reducing waste and optimizing the use of materials, reducing dependence on natural resources and environmental impact. Adopting circular economy strategies allows companies to be more sustainable, responsible and competitive. The main strategies for an effective transition to this model are detailed below.

Product Design for Circularity

Product design is critical to ensuring that goods can be easily recycled, reused, or repaired. Circularity begins at the design stage, where decisions are made that determine the life cycle of products and their environmental impact.

- **Use of Recyclable and Renewable Materials**: Companies should prioritize the use of recyclable and renewable materials in the design of their products. This includes the use of recycled metals, biodegradable plastics, and sustainable fibers, such as organic cotton or bamboo.

- **Modular Design**: Modular products make it easy to repair and replace individual components, extending their life and avoiding wasted materials. This approach is common in the tech industry, where some companies design electronic devices with interchangeable parts.

- **Avoid Toxic Substances**: The elimination of toxic substances in the manufacture of products is essential so that they can be recycled without risk of contamination. Companies must adhere to environmental safety standards and limit the use of hazardous chemicals.

- **Life Cycle Optimization**: Designing products that can be easily disassembled and sorted at the end of their useful life facilitates their recycling and reuse. This allows materials to be reincorporated into the production chain without losing quality.

Designing products for circularity is the first step towards a circular economy, as it allows companies to create sustainable goods that generate less waste and have a reduced environmental impact.

Implementation of Rental and Buyback Models

Renting and buying back products are effective strategies to maximize the use of goods and reduce the consumption of new resources. These models allow products to have a longer shelf life and help companies close the loop on their products.

- **Rental Model for Durable Products**: Renting products, such as furniture, vehicles, or appliances, allows consumers to access quality goods without needing to buy them. This model prolongs the use of the products and facilitates their maintenance and updating by the company.

- **Buyback Programs for Resale and Recycling**: Companies that offer buyback programs incentivize customers to return used products for resale or recycling. Patagonia, for example, buys back used clothes and resells them at discounted prices, encouraging reuse and preventing these garments from ending up in landfills.

- **Subscription Economy**: Some products are offered on a subscription basis, in which customers pay a fee to access services and goods on a temporary basis. This model allows products to be returned to the manufacturer to be refurbished or redistributed, reducing waste and generating recurring revenue for the company.

- **Advantages of Rental and Buyback Models for the Customer**: These models also offer advantages for the consumer, such as reduced costs, access to high-quality goods and the flexibility to change products when necessary.

The implementation of rental and buyback models allows companies to give a second life to their products, improving their efficiency and reducing environmental impact.

Reuse and Recycling of Materials in Production

Recycling and reusing materials in production processes are key strategies to reduce the use of raw materials and minimize waste.

- **Internal Recycling of Materials**: Companies can recycle waste generated in production to reuse it in new processes. This reduces the cost of materials and minimizes environmental impact. For example, in the fashion industry, some companies recycle fabric scraps to create new products.

- **Use of Recycled Materials in New Products**: By using recycled materials, companies can reduce their reliance on virgin raw materials. This includes the use of recycled plastics in packaging, recycled glass in bottles, or recycled aluminum in the manufacture of vehicles and other products.

- **Waste Collection Systems and Recoverable Waste**: Implementing selective waste collection systems in the facilities allows the separation of recoverable materials, such as paper, cardboard and metal.

These materials can be sold to recycling plants or reincorporated into production.

- **Collaboration with Suppliers of Recycled Materials**: Collaboration with suppliers that use recycled or sustainable materials is essential to implement a circular economy. Companies can seek alliances with responsible suppliers who supply them with materials with low environmental impact.

Recycling and reusing materials allows companies to reduce resource use, minimize waste, and promote a more sustainable production cycle.

Promoting Repair and Reuse

Encouraging the repair and reuse of products is one of the most effective strategies to extend their useful life and reduce the consumption of new resources. By encouraging consumers to repair their goods instead of throwing them away, companies promote a circular and sustainable economy.

- **In-House Repair Services**: Some companies, such as Patagonia, offer repair services for their products, allowing customers to extend their lifespan. These services may include repairs to clothing, footwear, electronics, and other consumer products.

- **Ease of Repair**: Designing products that make them easy to repair, such as the use of modular parts and easily accessible spare parts, allows customers to fix goods themselves or at local repair services. This approach reduces the cost and complexity of repair.

- **Consumer Repair Education**: Companies can offer tutorials and guides to teach consumers how to repair their products. This allows customers to gain skills in repairing their goods and reduces the amount of products that end up in landfills.

- **Second Life for Refurbished Products**: By refurbishing used products, companies can resell them at discounted prices, promoting reuse and offering affordable options for consumers. Refurbished products often undergo quality checks and repairs to ensure they are in good condition.

Promoting repair and reuse allows companies to strengthen their relationship with customers, reduce their environmental footprint and contribute to a more responsible consumption model.

Responsible Waste Management and Circular Economy Certifications

Responsible waste management and obtaining circular economy certifications are key tools for companies to demonstrate their commitment to sustainability and good practices.

- **Waste Management in the Production Process**: Implementing waste management policies from production allows waste to be reduced, classified and treated appropriately. Companies can apply practices such as composting organic waste or sorting hazardous waste for specialized treatment.

- **Circular Economy Certifications**: Circular economy certifications, such as Cradle to Cradle, ISO 14001, and the Global Circular Economy Standard, endorse the company's sustainable practices and give it credibility in the marketplace. These certifications demonstrate that the company complies with internationally recognized environmental and social standards.

- **Post-Consumer Waste Collection and Management Systems**: Companies can implement systems to collect end-of-life products and ensure their recycling or reuse. This includes collection programs for containers, batteries and electronic devices, facilitating their responsible disposal.

- **Transparency in Waste Management**: Informing consumers about how the company manages its waste and the efforts it makes to reduce it promotes transparency and strengthens trust in the brand. Companies can include this information in their sustainability reports and in their communications with customers.

Waste management and certifications allow companies to ensure proper treatment of their waste and demonstrate their commitment to the circular economy.

Adopting strategies for the circular economy allows companies to make the most of their resources, reduce their dependence on raw materials and minimise their environmental impact. From product design to waste management, every step towards circularity contributes to a more sustainable and profitable business model. The circular economy not only helps to reduce waste, but also generates added value for consumers and improves the reputation of companies in an increasingly conscious market. Implementing these strategies is essential for organizations to remain competitive and responsible in an environment that demands more and more commitment to caring for the planet.

Transparency and CSR

Corporate Social Responsibility (CSR) is a company's commitment to society and the environment, beyond its economic function. Transparency and accountability in these practices are essential to build trust with stakeholders and strengthen the company's reputation. By implementing CSR policies and adopting sustainability standards, companies can show their commitment to social and environmental well-being and position themselves as responsible leaders in their industries. Below are the key strategies to promote transparency and accountability in the company.

Publication of Sustainability Reports

Sustainability reporting is a critical tool for communicating the company's progress in its CSR initiatives and sustainable practices. These reports offer a comprehensive view of the organization's environmental, social, and economic impacts.

- **Transparency in Energy Consumption and CO_2 Emissions**: Reporting on energy consumption and greenhouse gas emissions allows companies to show their efforts to reduce their carbon footprint. This type of data gives stakeholders a clear view of the actions the company is taking to mitigate climate change.

- **Waste and Resource Management**: Reports should also include data on waste management, recycling, and the efficient use of natural resources. By demonstrating how materials are managed in the production chain, companies can show their commitment to the circular economy and minimizing waste.

- **Social Impact Indicators**: Sustainability reports should cover social indicators, such as diversity and inclusion, working conditions, and contributions to local communities. Showing progress in these issues helps companies build an image of responsibility and care for people.

- **Future Initiatives and Goals**: Beyond current achievements, reports should include the company's future sustainability goals and commitments. This allows stakeholders to learn about the company's direction and its ongoing commitment to improvement.

Publishing sustainability reports on a regular basis not only reinforces transparency, but also allows the company to identify areas for improvement and be held accountable for its CSR practices.

International Certifications and Standards

Sustainability certifications and international standards are a way to ensure that the company's practices comply with globally recognized requirements. Obtaining these certifications not only increases the company's credibility, but also allows it to differentiate itself in an increasingly sustainability-oriented market.

- **B Corp Certification**: This certification evaluates companies in terms of their social and environmental impact, transparency, and accountability. Obtaining B Corp certification is a seal of quality that demonstrates that the company operates for the benefit of the planet and society.

- **ISO 14001 for Environmental Management**: ISO 14001 establishes a framework for companies to manage their environmental impacts and improve their sustainable performance. By obtaining this certification, companies show their commitment to the responsible management of resources and the reduction of their environmental footprint.

- **Global Reporting Initiative (GRI) Standards**: GRI standards are an internationally recognized framework for sustainability reporting. Using GRIs ensures that the company's sustainability reporting meets transparent, comparable, and comprehensive criteria.

- **Cradle to Cradle Certification**: This certification assesses the life cycle of products and ensures that they can be reused safely and efficiently. The Cradle to Cradle certification guarantees that the company's products comply with the principles of the circular economy.

By earning these certifications and standards, companies demonstrate their commitment to sustainable practices and gain credibility among their customers and business partners.

Commitment to the Sustainable Development Goals (SDGs)

The UN's Sustainable Development Goals (SDGs) are a global roadmap for sustainable development. Aligning business activities with these goals allows companies to contribute to global sustainability goals.

- **Integrating the SDGs into Corporate Strategy**: Companies can review their strategy and operations to identify how they can contribute to the SDGs, such as the "Affordable and Clean Energy" goal or the "Climate Action" goal. Integrating the SDGs into the strategy allows sustainability to be a fundamental part of the business model.

- **Partnership for Gender Equality and Decent Work**: The goals of gender equality and decent work are fundamental to social sustainability. Companies can implement inclusion, equity, and fair working conditions policies to meet these goals and improve the work environment.

- **Promoting Innovation and Sustainable Infrastructure**: Companies can also align with the SDGs in terms of innovation and sustainability. Investing in clean technologies and sustainable infrastructure is one way to make progress towards meeting the Sustainable Development Goals.

- **Commitment to Climate Action and Ecosystem Protection**: Companies can take action to mitigate their environmental impact and contribute to climate action and protection of terrestrial and marine life goals. This can include reducing emissions, conserving ecosystems, and using resources responsibly.

Aligning business activities with the SDGs allows companies to collaborate in creating a more sustainable future and position themselves as responsible actors in the global community.

Participation in Sustainable Initiatives and Partnerships

Sustainable initiatives and partnerships are an effective way to drive collaborative change in the field of sustainability. Participating in these initiatives allows companies to learn from other actors, share experiences and achieve common goals.

- **United Nations Global Compact**: This initiative invites companies to commit to universal principles in human rights, labor standards, the environment, and anti-corruption. Participating in the Global Compact is a statement of commitment to sustainability and business ethics.

- **Partnerships with NGOs and Environmental Organizations**: Collaborating with non-governmental organizations and environmental entities allows companies to implement social responsibility programs, such as ecosystem conservation or the development of local communities. These alliances also provide companies with access to specialized knowledge and resources.

- **Energy and Clean Technology Innovation Programs**: Participating in technology innovation and renewable energy programs allows companies to be at the forefront of sustainability. Collaborative projects, such as clean energy research platforms, accelerate the transition to sustainable technologies.

- **Circular Economy Collaborative Networks**: Circular economy networks, which include businesses, governments, and organizations, facilitate the exchange of good practices and the creation of sustainable solutions. These collaborations promote the reuse and recycling of materials throughout the supply chain.

Participation in sustainable initiatives and alliances allows companies to act together to address global challenges and strengthen their position in the market as leaders in social and environmental responsibility.

Transparency and corporate social responsibility are essential pillars for companies to demonstrate their commitment to sustainability. Through sustainability reporting, international certifications, alignment with the SDGs, and participation in sustainable initiatives, companies can build a strong and trustworthy reputation. These practices allow organizations to operate ethically and responsibly, generating a positive impact on both the environment and society. By promoting transparency and accountability, companies not only contribute to more sustainable development, but also strengthen their position in a market that increasingly values commitment to the well-being of the planet and people.

Business transformation towards sustainability is a process that requires redefining success, implementing best practices, educating employees and adopting circular economy models. Companies that lead this change not only generate a positive impact on the environment and society, but also strengthen their reputation and prepare for a more sustainable future. Through transparency and corporate social responsibility, organizations can inspire other actors and contribute to building a global economy that respects and protects the planet.

EDUCATION AND ENVIRONMENTAL AWARENESS

Education and environmental awareness are essential to foster a society committed to sustainability. Educating people, especially new generations, about the importance of protecting the planet is key to lasting change. From educational programs to awareness campaigns, every effort contributes to empowering society to make informed and responsible decisions. Below, we explore the main strategies and examples that demonstrate how education and information drive environmental change.

Educating for Lasting Change

Educating in sustainability goes beyond transmitting theoretical knowledge; It also involves instilling values, skills, and attitudes that have a positive and lasting impact. Environmental education allows new generations to understand their relationship with the planet and enables them to face ecological challenges with a responsible and committed approach. Below are the main areas where sustainability education can make a significant difference.

Importance of Including Sustainability in Curricula

Integrating sustainability issues into the school curriculum fosters a deep understanding of the importance of the environment in all aspects of life. By incorporating these topics into early educational stages, students develop a solid foundation of knowledge and prepare to act consciously in their everyday lives.

- **Addressing Key Sustainability Themes**: Including topics such as climate change, biodiversity loss, circular economy, and renewable energy helps students understand how their decisions impact the planet. These topics can be adapted to different educational levels, from the most basic to the most advanced, facilitating progressive learning.

- **Promotion of Interdisciplinarity**: Sustainability encompasses several areas of knowledge, which allows environmental issues to be integrated into various subjects, such as science, economics, geography and ethics. This interdisciplinarity gives students a well-rounded view of how environmental issues are connected to other social and economic aspects.

- **Incorporating Classroom Projects and Practices**: Developing school projects that involve hands-on activities, such as recycling, tree planting, or building school gardens, reinforces theoretical learning and allows students to apply what they have learned in a real-world context.

Including sustainability in curricula allows students to grow up with a conscious and responsible mindset towards the environment, laying the foundations for a more informed and engaged society.

Promotion of Values of Responsibility and Respect for the Environment

Environmental education is also a tool to instill fundamental values, such as respect, responsibility and empathy, which motivate people to act for the benefit of the planet.

- **Develop Attitudes of Personal and Collective Responsibility**: Teaching students that their actions have an impact on both an individual and collective level is key to fostering responsibility. This teaching helps them understand that their choices, even if they seem small, contribute to a bigger change.

- **Promoting Respect for All Living Things**: Environmental education emphasizes respect for nature and living things, reminding students that all beings have a role in the balance of the ecosystem. This value is essential to promote a respectful attitude towards fauna, flora and natural resources.

- **Fostering Empathy and Commitment to the Common Good**: Empathy is a skill that can be developed through environmental education. By understanding the effects of the climate crisis and other environmental issues on vulnerable communities, students can develop a commitment to the common good and a motivation to protect the environment.

- **Developing Environmental Ethics in Decision-Making**: Environmental education also helps students understand the ethics behind their decisions. This means choosing responsible products, avoiding waste and opting for more sustainable consumption, based on values that prioritise respect for the planet.

Fostering these values from an early age allows students to develop an environmental awareness that will guide their decisions and behaviors throughout their lives.

Education in Skills for Sustainability

Sustainability depends not only on values, but also on practical skills that allow people to adopt a more environmentally friendly lifestyle. These skills are essential for students to be able to apply environmental knowledge in their everyday lives.

- **Recycling and Waste Management Skills**: Teaching students about waste separation and recycling allows them to reduce their ecological footprint. Knowing the recycling process, types of materials, and how to avoid using single-use plastics are practical and effective skills to minimize environmental impact.

- **Energy Efficiency and Resource Saving**: Instructing students on the efficient use of energy, water, and other natural resources is a valuable skill. Learning how to turn off lights, reduce water consumption, and opt for energy-efficient products are practices they can incorporate into their homes and future workplaces.

- **Responsible and Sustainable Consumption**: Education in responsible consumption teaches students to make informed and sustainable choices. This includes choosing local products, minimizing the use of plastic packaging, and considering the durability of goods, which fosters a circular economy and a more conscious lifestyle.

- **Solving Environmental Problems in Daily Life**: Developing skills to identify environmental problems and propose practical solutions, such as reducing food waste or creating home gardens, allows students to be agents of change in their communities.

These skills are tools that enable students to lead a sustainable life and actively contribute to environmental protection.

Early Awareness and Long-Term Impact

Early environmental education has a significant impact on the formation of responsible citizens committed to sustainability. Instilling these values and knowledge from childhood is crucial to establishing lasting change and an engaged mindset.

- **Benefits of Environmental Education from Childhood**: Studies show that children who receive environmental education from an early age are more likely to develop a lasting commitment to sustainability. By learning about the value of nature and the effects of climate change, children become aware and responsible adults.

- **Establishing Sustainable Habits and Routines**: Early awareness facilitates the creation of sustainable habits, such as saving water, reducing waste, and making healthy and sustainable food choices. These habits are incorporated into daily life and form a solid foundation for environmental behavior in the future.

- **Developing a Proactive Mindset Toward Environmental Issues**: Students who receive early environmental education are more likely to act proactively in the face of environmental issues. By understanding the challenges, they are motivated to seek solutions and lead initiatives that promote positive change.

- **Family and Community Involvement**: Environmental education in childhood also encourages the participation of families and communities. When children learn about sustainability, they are likely to share their knowledge with their families, which generates a multiplier effect on the environmental awareness of their immediate environment.

Early awareness is essential to ensure that sustainability is a value that transcends generations, creating citizens committed to the well-being of the planet.

Educating in sustainability is a process that goes beyond the transmission of knowledge; It involves fostering values, skills, and attitudes that have a positive and lasting impact on society. From the inclusion of sustainability in curricula to the promotion of practical skills and ethical values, environmental education lays the foundation for a change in mindset that can influence individual and collective behaviour. Environmental education is the key to forming an informed and committed society that works together to protect the planet, ensuring a more respectful and balanced future for future generations.

Educational Programs and Projects

Educational programs and projects are effective tools to promote environmental awareness and provide practical experiences that strengthen commitment to sustainability. By engaging students, communities, and businesses, these initiatives foster a direct connection to the environment and empower participants to take action in their daily lives and surroundings. Below are some of the most effective educational programs and projects and their benefits.

Eco-schools and Environmental Education Programs in Schools

Eco-schools are an example of an educational program that seeks to integrate sustainability into the school environment, promoting environmental responsibility from an early age.

- **Promoting Sustainable Practices in the School Environment**: In an Eco-school, students participate in activities such as recycling, waste reduction, water management, and energy saving. These practices teach them to apply sustainable principles in their day-to-day lives and to recognize the importance of reducing their ecological footprint.

- **Creation of Environmental Projects by Students**: Eco-schools often encourage students to create environmental projects in their schools, such as school gardens, recycling stations, and pollinator gardens. These projects encourage creativity and teamwork, while teaching practical skills and knowledge about caring for the environment.

- **Participation in Decision-Making**: In Eco-schools, students are encouraged to participate in the decision-making process about sustainability activities, which helps them develop a sense of responsibility and leadership. Involving students in the planning and execution of initiatives empowers them to act as agents of change in their community.

- **International Recognition and Support for Continuity**: Eco-schools have an international certification system that recognizes the school's commitment to sustainability. This certification encourages schools to maintain and improve their environmental practices, and motivates other institutions to join the program.

Eco-schools are a powerful tool for educating in sustainability and building an environmental culture from childhood, preparing students to act as future responsible leaders.

Reforestation and Conservation Projects

Participating in reforestation or conservation projects allows students and communities to experience first-hand the value of nature, while contributing to the restoration of local ecosystems.

- **Direct Connection to the Environment**: Reforestation allows participants to make a direct connection with nature. By planting trees and caring for ecosystems, students learn about biodiversity, the water cycle, and the importance of forests in carbon absorption and climate regulation.

- **Restoration and Maintenance of Natural Habitats**: The reforestation and conservation of natural areas allow habitats to be restored for local species, improving biodiversity and the health of ecosystems. These projects may include the recovery of native plant species and the control of invasive species that damage the natural balance.

- **Climate Change Awareness and Mitigation**: Reforestation is a key strategy to combat climate change, as trees absorb carbon dioxide. Participating in these projects allows students to understand the role nature plays in the fight against global warming and the importance of protecting natural resources.

- **Skills Development in Conservation and Ecology**: These projects provide participants with practical skills in ecology and conservation, such as species identification, monitoring ecosystem health, and maintaining green areas.

Reforestation and conservation projects offer an in-depth educational experience that sensitizes participants to the fundamental role of ecosystems in sustainability.

Workshops and Talks on Renewable Energies

Workshops and talks on renewable energies are an excellent way to educate the population about sustainable energy alternatives and their importance in reducing environmental impact.

- **Introduction to Clean Energy Sources**: These workshops educate participants about the various renewable energy sources, such as solar, wind, geothermal, and hydroelectric. Understanding how they work and their benefits over fossil fuels allows participants to make informed decisions about energy consumption.

- **Hands-on Demonstrations and Experiments**: Through demonstrations and experiments, such as building homemade solar panels or installing small wind generators, participants learn in practical ways how to harness renewable energy.

- **Energy Consumption and Efficiency Awareness**: The workshops also include tips on how to reduce energy consumption at home and at work, such as using efficient appliances and taking advantage of natural light. This knowledge allows participants to implement sustainable practices in their daily lives.

- **Knowledge about the Environmental Impact of Energy Sources**: Talks on renewable energy also explain the environmental impact of traditional energy sources, such as coal and oil. This helps participants understand why transitioning to clean energy sources is critical.

Workshops and talks on renewable energy provide participants with essential knowledge to promote a shift towards a more sustainable energy system.

Environmental Education and Outdoor Activities Centers

Environmental education centers offer educational experiences in direct contact with nature, which facilitates learning through observation and practice. Outdoor activities reinforce environmental awareness and respect for ecosystems.

- **Education in a Natural Environment**: Environmental education centers provide a natural environment where participants can observe local flora and fauna. This direct contact with nature generates an appreciation for the environment and an understanding of the value of biodiversity.

- **Fauna and Flora Observation Activities**: Activities such as bird watching, insect monitoring, and the study of native plants allow participants to learn about the biological diversity of their environment and learn about the interdependence of species.

- **Hiking and Conservation Workshops**: Hiking and conservation workshops, such as cleaning up natural areas and planting trees, help participants understand the benefits of protecting and maintaining ecosystems.

- **Education about Ecosystems and their Balance**: Environmental education centers offer educational programs that explain how ecosystems work and the importance of each element in their balance. This understanding helps participants see the relationship between the health of the environment and human well-being.

The activities in the environmental education centers are transformative experiences that strengthen the bond with nature and promote the protection of ecosystems.

Community and Civic Engagement Programs

Citizen participation in environmental programs fosters collaboration among community members and promotes shared responsibility in protecting the environment.

- **Beach and Park Cleaning Days**: These days allow citizens to contribute directly to the cleaning and conservation of natural spaces. Waste collection helps reduce pollution and raises awareness among participants about the impact of waste on the environment.

- **Creating Community Gardens**: Community gardens encourage collaboration and teach sustainable practices, such as growing local food and composting organic waste. These gardens not only improve the quality of life of the participants, but also create a sense of belonging and respect for the land.

- **Recycling and Waste Reduction Programs**: Community-based recycling and waste reduction programs, which include the separation and collection of recyclable materials, educate citizens about the importance of the circular economy and waste reduction.

- **Awareness of Local Biodiversity and Habitat Conservation**: Programs that promote the conservation of local species and habitats, such as the protection of wetlands or the creation of biological corridors, encourage citizen participation in the conservation of the region's biodiversity.

- **Responsible Consumption and Efficient Use of Resources Campaigns**: Responsible consumption campaigns promote sustainable habits, such as reducing the use of plastics and saving water. These campaigns are an effective way to raise community awareness of the need to adopt sustainable practices.

Citizen engagement programs empower communities to work together to protect the environment, strengthening the collective commitment to sustainability.

Educational programs and projects create practical opportunities for individuals and communities to get involved in protecting the environment. From Eco-schools and environmental education centers to citizen engagement programs, these initiatives offer enriching experiences that strengthen the connection to the environment and promote the adoption of sustainable practices. By promoting environmental education in various contexts, a more conscious and committed society is built, prepared to face environmental challenges with responsibility and action.

Information Drives Action

Information is one of the most powerful engines for motivating environmental action. The media, social media and awareness campaigns play a key role in disseminating information about environmental issues and raising society's awareness of sustainability. By accessing accurate and timely information, people better understand the magnitude of ecological challenges and feel compelled to take action to protect the planet. Below are the main communication channels and strategies that drive the change towards a more sustainable lifestyle.

The Role of the Media in Raising Awareness

The media are key to informing and sensitizing public opinion about environmental problems. Through reports, documentaries and news, the media can significantly influence the perception and understanding of environmental issues.

- **Dissemination of Information on Global Environmental Issues**: The media report on topics such as climate change, biodiversity loss, and pollution, providing the public with a comprehensive view of the challenges globally. This information helps people understand the global context and its connection to local issues.

- **Investigative and In-Depth Reporting**: Investigative reporting provides a detailed look at environmental problems and their causes. These reports often include interviews with experts, scientific data, and testimonies from affected communities, providing an in-depth understanding of the problems and the urgency of solving them.

- **Creating Impactful Narratives to Generate Empathy**: Media has the ability to create stories that touch the audience emotionally, promoting empathy and understanding of the situation of communities affected by the environmental crisis. These narratives reinforce a sense of responsibility and commitment to change.

- **Promotion of Collective Action**: Through the dissemination of events, marches and campaigns, the media encourage citizens to participate in collective initiatives. Coverage of social movements, such as climate strikes, inspires more people to join the fight for sustainability.

The media, with its reach and influence, play a central role in the creation of an informed and aware society, capable of acting responsibly towards the environment.

Awareness Campaigns to Encourage Habit Change

Awareness campaigns, both locally and globally, are an effective tool to promote changes in behaviors and encourage sustainable practices. These campaigns use clear and engaging messages to inspire people to take action.

- **Global Initiatives such as "Earth Hour"**: Campaigns such as "Earth Hour", in which millions of people turn off their lights for an hour, generate a significant impact on raising awareness about energy consumption. These massive events create a sense of global community and remind society of the importance of taking sustainable action.

- **Waste Reduction Movements, such as "Zero Waste"**: "Zero Waste" campaigns promote a zero-waste lifestyle, encouraging people to reduce the use of single-use plastics and minimize waste. These campaigns encourage responsible consumption and contribute to reducing pollution.

- **Local Plastic Reduction and Water Saving Campaigns**: Local campaigns, such as plastic reduction and responsible water use campaigns, address community-specific issues. These initiatives often involve local organizations and allow for a closer connection with the population, motivating concrete and tangible changes in lifestyle.

- **Spreading Good Environmental Habits Through Social Media**: Social media campaigns offer sustainable tips and practices, such as recycling and saving energy. These types of campaigns reach a wide audience and are easy to share, making it easier to spread good habits in the community.

Awareness campaigns show that by joining forces, every person can contribute to change. These initiatives inspire society to adopt sustainable practices and demonstrate the positive impact of small actions.

Digital Education and Online Resources

Technology and digital platforms have expanded access to environmental information, allowing anyone, anywhere, to learn about sustainability and act accordingly. Online resources have become key tools for educating society about environmental problems and available solutions.

- **Environmental Awareness Documentaries and Series**: Documentaries and series on environmental issues, such as "Our Planet" or "Before the Flood," provide visual and engaging information about environmental challenges. This content raises awareness by showing the effects of climate change and biodiversity loss in a shocking way.

- **Virtual Sustainability Courses and Workshops**: There are numerous online courses on sustainability, recycling, renewable energy, and circular economy. These courses allow people to learn about sustainability at their own pace and gain practical knowledge that they can apply in their daily lives.

- **Blogs and Educational Resource Websites**: Blogs and websites dedicated to sustainability, such as Treehugger or GreenPeace, offer articles, guides, and tips on sustainable practices. These resources are easily accessible and provide useful, up-to-date information for people of all ages.

- **Consumption and Emissions Monitoring Apps**: There are mobile apps that allow users to monitor their energy consumption, calculate their carbon footprint, and receive recommendations to reduce their environmental impact. These apps are practical tools that help users lead a more sustainable lifestyle.

Online resources provide accessible and ongoing education on sustainability, empowering people to become agents of change in their communities.

Opinion Leaders in Sustainability

Influencers and thought leaders in the field of sustainability have a significant influence on creating environmental awareness and promoting sustainable practices. With their platforms, they inspire thousands or even millions of people to adopt a more conscious lifestyle.

- **Spreading Sustainable Messages on a Large Scale**: Influencers have a massive reach on social media, allowing them to reach a wide audience with their sustainability messages. By sharing content about responsible consumption, recycling and energy saving, these opinion leaders motivate their followers to change their habits.

- **Collaboration with Sustainable Brands and Companies**: Many influencers collaborate with brands committed to sustainability, promoting responsible products and services. These collaborations increase the visibility of sustainable brands and show consumers environmentally friendly options.

- **Inspiration and Example of Good Practices**: By sharing their own sustainable lifestyle, influencers serve as an example for their followers. By showing how they integrate responsible practices into their everyday lives, they inspire others to do the same.

- **Creating Communities of Environmental Awareness**: Sustainability influencers often create communities where their followers can share ideas and experiences about responsible practices. These communities strengthen the engagement of participants and facilitate the exchange of information on sustainability.

Influencers and opinion leaders contribute to a more informed and engaged society, showing that sustainability can be part of everyday life and creating a multiplier effect in the adoption of responsible practices.

Transparency and Access to Environmental Information

Transparency in the disclosure of environmental data is essential for society to be able to make informed decisions. By having access to information about the environmental impact of companies and governments, citizens can demand change and promote more sustainable practices.

- **Publication of Data on CO_2 Emissions and Resource Consumption**: Companies and governments that publish their CO_2 emissions and resource consumption allow citizens to know their environmental impact. This information also drives companies to improve their practices to reduce their ecological footprint.

- **Access to Air and Water Quality Information**: Transparency in air and water quality is critical to protecting public health. By accessing this data, people can take precautions and demand improvement measures in areas with high levels of pollution.

- **Corporate Sustainability and CSR** Reporting: Companies that publish sustainability and corporate social responsibility (CSR) reports show their commitment to transparency and enable consumers to make informed decisions. These reports detail the company's actions on issues such as energy consumption, recycling and biodiversity protection.

- **Biodiversity and Ecosystem Maps and Reports**: Access to data on biodiversity and ecosystems allows citizens to understand the state of nature in their region. This information is key to promoting conservation and to making informed decisions in development and urbanization projects.

Transparency and access to environmental information are essential to create an informed and empowered society, capable of acting for the benefit of the planet.

Information and transparency are powerful tools to motivate environmental action and build a more conscious and engaged society. Through media, awareness campaigns, digital resources, and data transparency, every individual can access knowledge that fosters sustainable practices. By understanding the magnitude of environmental challenges, people are better prepared to make responsible decisions and contribute to the protection of the planet, demonstrating that knowledge is the first step towards change.

Fostering Professional Sustainability

Environmental education is a fundamental pillar in the training of professionals who not only understand the importance of sustainability, but are also equipped to implement significant changes in their respective industries. As the demand for sustainable practices grows, it is crucial that future leaders and experts are adequately prepared to face environmental challenges in innovative and responsible ways. The key aspects of the role of environmental education in the training of these professionals are detailed below.

Mainstreaming Sustainability in Higher Education

More and more higher education institutions are integrating sustainability into their academic programs, reflecting the growing importance of these issues in the world of work and in society at large.

- **Specialized Programs in Sustainability**: Universities and training centers are developing specific programs and careers in areas such as environmental sciences, natural resource management, environmental engineering, and sustainable architecture. These programs prepare students with technical and practical knowledge to address contemporary environmental issues.

- **Interdisciplinarity in Training**: Sustainability is a topic that spans multiple disciplines, so it is essential to encourage an interdisciplinary approach. Higher education programs can include subjects in social sciences, economics, and technology along with environmental topics, providing students with a holistic understanding of challenges and opportunities in sustainability.

- **Applied Research and Practical Projects**: Universities can encourage applied research in sustainability by encouraging students to participate in projects that address real environmental problems. These projects allow students to apply their knowledge in practical contexts, developing critical skills and innovative solutions.

- **Collaboration with Companies and Organizations**: Partnerships between universities and companies can provide students with internship and internship opportunities in sustainability contexts. This practical experience is vital to develop skills in environmental management and the implementation of sustainable practices in the workplace.

Incorporating sustainability into higher education is essential to prepare professionals who will lead the shift towards a more sustainable future.

Business Training for Sustainability

Companies play a crucial role in implementing sustainable practices, and training their employees in this area is critical to making a real impact.

- **Sustainability Training Programs**: Companies can develop training programs that address key topics such as energy efficiency, waste management, responsible use of resources, and sustainability in the supply chain. These trainings help employees understand the importance of their actions and apply sustainable practices in their daily work.

- **Development of Technical and Practical Skills**: Trainings should include specific technical skills, such as the use of tools for carbon footprint analysis, the implementation of recycling systems, and the adoption of clean technologies. By developing these skills, employees can actively contribute to the company's sustainability goals.

- **Fostering a Sustainable Organizational Culture**: Sustainability training must also be accompanied by a cultural change in the organization. Fostering an environment that values sustainability and promotes collaboration among employees contributes to the effective implementation of learned practices.

- **Evaluation and Continuous Improvement**: Companies must establish mechanisms to evaluate the impact of training on sustainability and make continuous improvements in their programs. This includes tracking sustainability indicators and employee feedback on the practices implemented.

Sustainability training within companies is critical to ensure that employees are prepared to adopt responsible practices and contribute to change.

Network of Professionals in Sustainability and Good Practices

The creation of networks and associations of sustainability professionals is key to sharing knowledge, experiences and good practices.

- **Fostering Collaboration Between Professionals**: Professional networks allow individuals to connect with others who share similar interests in sustainability. These connections foster collaboration in projects, research and development of innovative solutions to environmental problems.

- **Exchange of Good Practices**: Sustainability associations can organize conferences, workshops, and seminars where professionals share experiences and successful practices. This exchange of good practices is essential for learning and continuous improvement in the field of sustainability.

- **Access to Resources and Training Opportunities**: Through these networks, professionals can access educational resources, training courses, and mentoring opportunities. This contributes to ongoing professional development and strengthening leadership capacity in sustainability.

- **Development of Collaborative Projects**: Networks allow the creation of collaborative projects between different sectors, such as companies, non-governmental organizations and academics. These projects can address specific sustainability challenges and foster greater awareness and collective action.

Networks of sustainability professionals are essential to drive collaboration and knowledge sharing, strengthening commitment to sustainability in various industries.

Promotion of Sustainable Entrepreneurship

Environmental education can also inspire young entrepreneurs to develop initiatives that promote sustainability and respond to today's environmental challenges.

- **Encourage Entrepreneurship in Sustainable Sectors**: Educational programs can foster interest in sustainable entrepreneurship by offering training in business management with a focus on sustainability. This includes topics such as the circular economy, organic agriculture and renewable energy.

- **Creation of Sustainable Business Incubators and Accelerators**: Incubators and accelerators can offer support to entrepreneurs who develop sustainable business ideas. These programs can include mentoring, seed funding, and training in sustainable practices, helping entrepreneurs turn their ideas into viable ventures.

- **Developing Innovative Solutions to Environmental Challenges**: Sustainable entrepreneurs are in a unique position to develop innovative solutions that address environmental problems. This can include clean technologies, eco-friendly products, and services that reduce environmental impact.

- **Promoting Sustainable Entrepreneur Networks**: Networks that connect sustainable entrepreneurs foster collaboration and the exchange of ideas. These networks can help entrepreneurs learn from each other, find investors, and access market opportunities.

Promoting sustainable entrepreneurship not only contributes to the economy, but also addresses environmental challenges, generating a positive impact on society and the environment.

Environmental education is essential to train professionals who will lead the change towards a more sustainable future. From incorporating sustainability into higher education to in-company training and professional networking, every aspect contributes to empowering people to act responsibly and effectively in their respective fields. By inspiring young entrepreneurs and promoting collaboration in sustainability, the foundations are laid for a more conscious economy and a positive impact on the environment and society. Training sustainability professionals is a crucial step towards building a more planet-friendly future.

Environmental Responsibility in the Community

Environmental education should not be limited to educational institutions and companies; it is essential that it be extended to the community as a whole. Creating a culture of environmental responsibility allows individuals, families, and neighbors to work together to protect and conserve their local environment. Fostering this collective commitment not only improves quality of life, but also empowers people to take action for sustainability. Below, various strategies for cultivating this culture in the community are explored.

Community Awareness Events

Community events are a great way to bring people together and provide them with information about sustainability in an accessible and practical way.

- **Environmental Fairs**: Hosting environmental fairs allows community members to learn about sustainable practices through workshops, exhibitions, and interactive activities. These events can include activities for children, recycling demonstrations, and exhibits on renewable energy.

- **Talks and Conferences**: Inviting sustainability experts to give talks or lectures on relevant environmental topics helps increase awareness of local issues, such as pollution, resource conservation, and biodiversity. These presentations can motivate attendees to take a more proactive approach to the environment.

- **Hands-on Workshops**: Workshops on composting, organic gardening, recycling techniques, and energy saving offer participants practical tools that they can implement in their homes. These hands-on experiences are effective in teaching people how to reduce their ecological footprint.

- **Networking activities**: These events also encourage networking among community members, facilitating collaboration on future projects and promoting a sense of belonging and shared responsibility towards the environment.

Hosting community awareness events creates a space for learning and interaction, where people can connect with their neighbors and take an active role in protecting the environment.

Green Spaces and Community Gardens

Community gardens and green spaces are vital to improving quality of life and educating the community about sustainability and biodiversity.

- **Community Garden Development**: Creating community gardens allows residents to grow their own food, learn about urban agriculture, and understand the life cycle of plants. These spaces encourage local food production and help reduce the carbon footprint associated with transporting products.

- **Biodiversity and Ecosystems Education**: Community gardens are ideal educational opportunities. Through gardening activities, participants learn about the importance of biodiversity, the water cycle, and pollination, promoting a deeper understanding of the environment.

- **Spaces for Recreation and Socialization**: Green spaces serve as areas for recreation and socialization, where communities can meet and share experiences. These environments foster social cohesion and collaboration between neighbours.

- **Local Reforestation Initiatives**: Reforestation projects in community areas help restore habitats and improve air quality. Involving the community in these initiatives generates a sense of pride and responsibility towards nature.

Creating green spaces and community gardens not only enhances the local environment, but also educates and unites community members around sustainability.

Ecosystem Cleanup and Restoration

Participation in clean-up and restoration projects is an effective way to raise community awareness about pollution and the importance of protecting ecosystems.

- **Cleaning of Rivers, Beaches and Parks**: Organizing clean-up days in natural spaces allows citizens to see first-hand the impact of waste on the environment. These activities promote collaboration and strengthen the collective commitment to conservation.

- **Habitat Restoration**: Habitat restoration projects, such as reforestation or the recovery of degraded areas, are vital to improving the health of local ecosystems. Involving the community in these initiatives fosters a sense of ownership and responsibility over their environment.

- **Waste Impact Education**: Through clean-up projects, participants learn about the relationship between their consumption habits and pollution. This hands-on education helps raise awareness of the need to reduce waste and adopt a more sustainable lifestyle.

- **Promoting Species Conservation**: Projects that involve the conservation of local species and the protection of critical ecosystems are opportunities to educate the community about biodiversity and the importance of protecting endangered species.

Ecosystem clean-up and restoration projects empower communities and give them the opportunity to take direct action in the conservation of their environment.

Responsible Consumption and Zero Waste

Responsible consumption and zero waste campaigns are effective in raising awareness in the community about the importance of adopting sustainable habits.

- **Education on Responsible Consumption**: Campaigns that promote responsible consumption address issues such as reducing the use of plastics, buying local products and choosing products with less environmental impact. Clear and accessible information can influence citizens' purchasing decisions.

- **Development of Zero Waste Initiatives**: Zero waste initiatives encourage the reduction, reuse and recycling of materials. These campaigns can include workshops on how to make homemade cleaning products, how to reuse materials or how to compost at home.

- **Creating Community Challenges**: Organizing community challenges to reduce the use of plastics or waste can motivate citizens to actively participate and share their experiences. Friendly competition can result in behavioral changes and increased awareness of sustainability.

- **Collaboration with Local Businesses**: Responsible consumption campaigns can include collaborations with local businesses that promote sustainable products. By partnering with these businesses, you support the local economy and foster positive change in the community.

Responsible consumption and zero waste campaigns inspire citizens to adopt more sustainable habits and become environmental advocates.

Promoting Clean Energy in Homes

Encouraging the use of clean energy and energy efficiency in homes is an important step in reducing the carbon footprint of families.

- **Installing Renewable Energy**: Promoting the installation of solar panels and other renewable energy sources in homes allows families to reduce their dependence on fossil fuels and lower their energy bills.

- **Use of Efficient Appliances**: Educating the community about the importance of using energy-efficient appliances contributes to the reduction of energy consumption in the home. Offering incentives or subsidies for the purchase of these devices can motivate more families to make changes.

- **Energy Efficiency Workshops**: Organizing workshops on how to improve energy efficiency in the home, such as properly insulating windows and doors, using LED lighting, and reducing energy consumption, can help citizens implement simple and effective changes.

- **Creating Clean Energy Groups**: Encouraging the creation of community groups focused on promoting clean energy can incentivize families to adopt sustainable practices, share resources, and collaborate on projects that encourage the use of renewable energy.

Promoting clean energy in homes not only benefits the environment, but also allows families to save money and improve their quality of life.

Creating a culture of environmental responsibility in the community is essential to the collective commitment to sustainability. Through the organization of awareness events, the creation of green spaces, participation in cleaning projects, responsible consumption campaigns and the promotion of clean energy, a more conscious and active society in environmental protection is built. This community engagement not only generates a positive impact on the local environment, but also inspires others to join the cause, creating a multiplier effect in the fight for a more sustainable future.

Education and environmental awareness are essential to empower society in the transition to a more sustainable future. From training young people to educating professionals and building engaged communities, every effort contributes to protecting the planet. By implementing educational programs, fostering transparency, and promoting a culture of sustainability, we can build a more informed and responsible society, capable of facing environmental challenges with knowledge and commitment. Environmental education is an investment in the future, ensuring a healthier, more balanced and fairer world for generations to come.

THE FUTURE IS IN OUR HANDS

In a world that faces climate change as one of the most significant challenges of our time, it is essential to remember that the future of our planet is in our hands. Every small step we take, every decision we make can have a profound impact on the environment around us and on the legacy we will leave to future generations. This is a crucial moment to reflect on the collective responsibility we share and to recognize that the fight against climate change is not just a challenge for governments or large corporations, but a call to action that is incumbent on each of us. It doesn't matter who we are, where we come from, or what our context is; we all share this home called Earth, and it is our obligation to care for and protect it.

As the planet's climate continues to change, the effects are becoming more evident and are being felt in every corner of the world. From devastating droughts that wipe out agricultural land to catastrophic floods that destroy communities, from rising temperatures that threaten the health of millions of people to the irreparable loss of biodiversity, these issues affect us all. However, it is the most vulnerable communities that suffer the most severe consequences, highlighting the urgent need for an equitable and fair approach to our solutions. This is a time to look around us, to acknowledge the suffering of those who are most affected, and to come together in the search for answers.

It is critical for each of us to understand that while our individual actions may seem small compared to the magnitude of the climate change problem, the sum of our efforts can bring about significant change. Every decision we make, from how we move around to what we consume, has the power to influence the environment. By choosing responsible drinking habits, participating in community initiatives, and educating those around us, we are contributing to a larger movement. The sum of small actions, from leaving the car at home and opting for cycling or public transport, to the simple act of taking a reusable bag to the supermarket, can lead to a wave of change.

Education and awareness are powerful tools in this fight. Reflecting on the role of environmental education is essential to empower future generations. By cultivating awareness about the impact of our decisions and actions, we are sowing the seeds of change in the minds and hearts of young people. It is critical to empower new generations to become environmental advocates, to understand the magnitude of the challenges we face and to be inspired to act. When we teach children and young people about the importance of taking care of our planet, we are giving them the necessary tools to build a more sustainable and just future.

Despite the gravity of the situation, we must never lose hope. Every day, innovations and sustainable practices emerge that prove that a better future is possible. There are inspiring stories of communities coming together to restore ecosystems, companies adopting sustainable business models, and individuals choosing to make the planet their priority. These stories are a testament to the power of collective action and the human ability to adapt and overcome challenges. They remind us that, despite adversity, there is always room for hope and transformation.

Recognizing that our actions today will impact future generations is a call to intergenerational responsibility. By committing to building a more sustainable future, we are securing a livable world for those who will come after us. This legacy is a treasure that we must cherish and protect. The urgency of our actions translates into an opportunity to innovate, to rediscover our relationship with the planet and to create a healthier and more resilient environment. Climate change is not only a crisis; It is a moment of reflection, an instant in which we can reassess our priorities and make sustainability a fundamental part of our lives.

This is a call to each of you, to every individual who is overwhelmed by the magnitude of the crisis, to every young person who dreams of a better future, and to every elderly person who has seen the world transform. The future is in our hands, and it is our responsibility to ensure that it is a future that reflects our values of care, respect, and love for the planet. Let us embrace this challenge with optimism and with the conviction that each of us can make a difference.

Let us invoke the spirit of collaboration and unity, because together we are stronger. May our fight be for a world where sustainability is the norm and not the exception, where every action counts and where every voice is heard. Let's move forward, hand in hand, towards a brighter and more sustainable future. Let's act now, because change is not only possible, it is our responsibility and our privilege.

The future is in our hands!

RECOMMENDED READING

- **"The Uninhabitable Earth: Life After Warming"** by David Wallace-WellsA shocking analysis of the potential consequences of climate change if we don't act immediately.

- **"This Changes Everything: Capitalism vs. the Climate"** by Naomi KleinA powerful critique of how the capitalist system is intrinsically linked to climate change and how we can transform our economy.

- **"The Sixth Extinction: An Unnatural History"** by Elizabeth Kolbert, an examination of biodiversity loss and how climate change is accelerating species extinction.

- **"Drawdown: The Most Comprehensive Plan Ever Proposed to Reverse Global Warming"** edited by Paul HawkenThis book presents practical and feasible solutions to address climate change.

- **"In Wild Company"** by Manuel RivasA novel that explores the connection between humans and nature, and the urgent need to take care of our planet.

- **"Field Notes from a Catastrophe: Man, Nature, and Climate Change"** by Elizabeth KolbertA field account showing the impact of climate change in different parts of the world.

- **"This Is How You Lose the Time War"** by Amal El-Mohtar and Max Gladstone, a science fiction novel that explores the connection to nature and the impact of our actions on time.

- **"Climate: A New Story"** by Charles EisensteinA deep reflection on how we need to change our narrative around climate and nature.

- **"Braiding Sweetgrass: Indigenous Wisdom, Scientific Knowledge, and the Teachings of Plants"** by Robin Wall Kimmerer, a work that intertwines indigenous wisdom with science, highlighting the relationship between people and nature.

- **"The Overstory"** by Richard PowersA novel that intertwines the lives of various characters with the lives of trees, exploring the interconnection between humans and nature.

- **"The Hidden Life of Trees: What They Feel, How They Communicate"** by Peter Wohlleben, a book that reveals the fascinating world of trees and their role in the ecosystem.

- **"The Limits to Growth"** by Donella Meadows et al. A classic study on population growth and sustainability in a finite world.

- **"No One Is Too Small to Make a Difference"** by Greta ThunbergA collection of speeches by the climate activist, inspiring people to act in the face of the climate crisis.